northwest

berry

cookbook

Finding,

Growing,

and Cooking

with Berries

Year-Round

Kathleen Desmond Stang

SASQUATCH BOOKS
SEATTLE

Printed in the United States of America.
Distributed in Canada by Raincoast Books Ltd.
02 01 00 99 98   5 4 3 2 1

Cover and interior design: Karen Schober
Cover photograph: Angie Norwood Browne
Illustrations: Tami Knight
Composition: Patrick Barber

Library of Congress Cataloging in Publication Data
        Stang, Kathleen Desmond.
                        Northwest berry cookbook : finding, growing, and cooking with berries year-round / Kathleen Desmond Stang.
                        p. cm.
                        Includes index.
                        ISBN 1-57061-112-2
                        1. Cookery (Berries) 2. Berries—Northwest, Pacific. I. Title.
        TX813.B4S73 1998
        641.6'47—dc21   98-10124

Sasquatch Books
615 Second Avenue
Seattle, Washington 98104
(206) 467-4300
books@sasquatchbooks.com
http://www.sasquatchbooks.com

Sasquatch Books publishes high-quality adult nonfiction and children's books related to the Northwest (Alaska to San Francisco). For more information about our titles, contact us at the address above, or view our site on the World Wide Web.

This book is dedicated to my friends and colleagues in the Pacific Northwest.

Cooks, chefs, and eaters.

Growers, processors, and providers.

Librarians, county extension agents, journalists, and friends.

And to Bob, the "taste tester."

I also appreciate the help I received from:

Berry Works, California Strawberry Commission, North American Blueberry

Council, Ocean Spray Cranberries, Oregon Raspberry and Blackberry Commission,

and the Washington Red Raspberry Commission.

Special thanks to

Gary Luke and Susan Derecskey. And to Chad Finn, Gary Melton, Kim Patten,

and Bill Peters—berry experts who kept me on the right track.

contents

recipe list

Breakfast and Breads

Salads and Entrees

Condiments and Preserves

recipe list

Cakes

Pies and Tarts

Other Desserts

Berries. Wonderful Berries.

I love berries. All kinds.

Tiny blueberries, gathered hiking in the Cascade Mountains and gobbled up on the spot.

Fragrant, totally ripe strawberries, tasting like the little ones I remember as a kid.

Glistening blackberries, baked in a pie.

Warm jewel-like cranberries, for the holidays.

———

Summer berries—strawberries, blueberries, and bramble berries—have similar characteristics. All are high in natural sugar, quite perishable, and can be served either fresh or cooked. These berries can generally be used interchangeably in recipes.

Currants and gooseberries are summer fruit too. But because of their high acidity, they seem more like lingonberries and tart cranberries.

This book features a dozen or so of the most widely available berries found in the Pacific Northwest, including Alaska, British Columbia, Washington, Oregon, and northern California.

The first berries to reach Northwest markets each year are strawberries from California, the harbinger of spring. In late spring, local strawberries start to arrive, with pink and green gooseberries not far behind.

By the Fourth of July, the farmers' markets are overflowing with blueberries, red and black currants, and raspberries as well as strawberries. Then a kaleidoscope of blackberry hybrids. And on and on.

Then, as the days grow shorter and the harvest winds down, local cranberries hit the market.

Here in the Northwest, berries are a year-round fruit.

the
berries

Strawberry

Alpine Strawberry

Strawberries and Alpine Strawberries

Doubtless God could have made a better berry,
but doubtless God never did.

—William Butler (circa 1600)

Recipes Using Strawberries
and Alpine Berries

Strawberries and Alpine Strawberries

Like many of our neighbors, we have a small strawberry patch in the backyard.

Sometimes in early spring, with nary a little white blossom on any plant, I succumb and buy berries at the supermarket. I'm lured by the faint, come-hither strawberry aroma, but often the texture is hard, the flavor, fleeting. And I resolve to wait for the local harvest.

By June, however, it's a race with the birds and squirrels to see who can get to the berries first. I'd say that most of these berries are gobbled in the garden. Every morning throughout the summer, there are a few to pick. Occasionally I'll hoard a couple of days' worth to slice on cereal or top a shortcake, but these berries don't seem to keep well. God must have meant them to be eaten on the spot.

Wild strawberries bring back memories. I remember watching my youngest sister, playing with her miniature tea set out in the back garden under the ferns. To accompany the "tea" she served just-picked wild strawberries, no bigger than a fingertip, to her dolly and to herself.

———

The Latin name for strawberry, *Fragaria*, comes from the word meaning "fragrant." And fragrant it is.

More than two thousand years ago, the Romans, and perhaps the Greeks, were acquainted with aromatic and flavorful strawberries that still exist today, in particular the tiny wood strawberry *Fragaria vesca*, called *fraises des bois* in French. In the sixteenth century, Europeans began to cultivate the small wood strawberry and the musky *Fragaria moschata* in gardens.

The big strawberries we know today were developed with the crossing of the small

Virginia or scarlet strawberry, *Fragaria virginiana*, and the large Chilean or beach straw-berry, *Fragaria chiloensis*. The result was a large, sweet fruit with a distinct fragrance, which soon became the standard against which all strawberries were measured.

The tale starts in 1712, when a French navy officer, Amedée François Frezier, traveled to Chile as a spy and an amateur botanist. Frezier (whose name, curiously enough, is derived from *fraisier*, meaning "strawberry plant" in French) found the Chilean strawberry, which had been cultivated in Chile for many years. He was im-pressed by the size of the fruit, "as big as a walnut," and took some home to France. Unfortunately, he brought only plants "heavy with fruit," not realizing he had selected only female specimens, which could not reproduce.

The actual development of the big-fruited strawberry, however, lay with An-toine Nicolas Duchesne, a young Frenchman who was familiar with the Virginia strawberry, which grew wild throughout eastern North America and had been known in Europe since about 1600. Duchesne completed the final step in 1765 by crossing the Virginia berry with Frezier's large Chilean berry, producing *Fragaria* × *ananassa*, the first cultivated variety using standard-size berries.

Strawberries

Strawberries grow in every state of the union. The vast majority of the com-mercial crop, more than 80 percent, comes from California, where the season be-gins in January and lasts well into November, depending on the weather. In effect, strawberries are available practically year-round. In the Northwest, however, local strawberries don't usually ripen until late May, at the earliest, or mid-June. But, as any old-timer will tell you, these berries are well worth waiting for.

First the names: Hood, Rainier, Shuksan. These berries are big and craggy like the mountain peaks they are named for. Then the fragrance. Just driving by a U-pick farm on a warm afternoon, there is an aroma like freshly made jam wafting on the wind. When these berries are ripe, they are sweet and flavorful. When picking them, it's a real temptation to pop more in your mouth than into your basket.

There is a minor downside to these exceptional Northwest berries, however. They are extremely fragile and need to be dealt with immediately. With really ripe berries, for example, it may be just a matter of hours before they are mush. Keep them cool and use them as soon as possible.

Alpine Strawberries

The tiny northern European *fraise des bois* (wood strawberry), *Fragaria vesca*, sometimes called alpine strawberry, is hundreds of years removed from the large-fruited kinds that have been developed since the mid-eighteenth century. And the little white, yellow, or red berries, with crunchy seeds and a fabulous fragrance, are hard to find. They are grown on a small scale in California and in the Northeast. Look for them on the menus of fancy restaurants or in little baskets at upscale produce markets, for a price. Or check local nurseries for plants or seeds.

Picking Strawberries

For best quality, pick strawberries when fully ripe and at their peak of flavor. Harvest early in the day, if possible, when it is cooler. Use shallow containers to collect the berries without crushing. Keep them cool in the shade and put them in the refrigerator, unrinsed, as soon as you get home.

Growing Strawberries in the Home Garden

Strawberries are one of the easiest fruits to grow in the home garden, and among the most productive. They require little room and are particularly well suited to the Northwest climate. The key is to match the cultivar to your location and climatic conditions.

Generally, strawberries are planted in the spring or early in the fall, so that the plants can become established before the cold weather sets in. The plants are usually sold bare root. Be sure your plants are certified virus-free and come from a reputable nursery or catalogue.

Plant them in a sunny area in well-drained soil, rich in organic matter. Raised beds help avoid root diseases and make it easier on your back with these low-growing plants. Avoid areas where tomatoes, peppers, potatoes, or other members of the *Solanaceous* family have recently grown, and avoid former bramble berry sites.

When planting, take care that the crown of the plant is placed neither too high, which would expose the roots and allow the plant to dry out, nor too low, which would cause the plant to rot. A straw mulch helps keep the fruit clean and dry and the soil moist. In areas with hard winters, the straw can protect the plants from injury from the cold.

Gratification comes quickly with strawberries. With day-neutral plants, you'll probably have berries in your first year.

Many of the strawberry cultivars in the Northwest (see pages 20-21) were developed at regional breeding programs in Oregon, Washington, British Columbia, and California. The three basic types of strawberries available in the Pacific Northwest are:

The June-bearing cultivars. Also called single cropping or short day, these produce fruit in June. The previous summer and fall, when the days become shorter, these plants develop the buds that will be blossoms the following spring, and ripe berries about a month later.

The day-neutral cultivars. These are a fairly recent development. They form their fruit buds without regard to the day length. These plants flower and bear fruit throughout the growing season (which can be up to five months) if they are well cared for and the weather isn't too hot. The plants are smaller and produce fewer runners, but give more fruit than either of the other two types.

The ever-bearing cultivars. These produce two crops each year, one in June and the other in the fall. The name "ever-bearing" is somewhat misleading, however, since they produce less fruit than the day-neutral cultivars. This may be why many gardeners are replacing their ever-bearers with the newer day-neutral types.

Alpine berries and tiny *fraises des bois* are available from nurseries in pots. Tuck them in the garden near where you sit, in order to enjoy their heady, sweet fragrance. Or plant them next to a walkway, so you can pick the ripe ones every day or two. The white varieties have an extra advantage—the birds peck only at the red ones.

GOOD STRAWBERRY CULTIVARS FOR THE PACIFIC NORTHWEST

DAY-NEUTRAL CULTIVARS			
CULTIVAR	**SEASON**	**EAST/WEST**	**CHARACTERISTICS**
Seascape	June–Oct.	West	Large, firm fruit with very good flavor; excellent fresh, very good for preserves and freezing; strong plants.
Selva	June–Oct.	West	Large, firm fruit, turns red before fully ripe, less juicy than most; good for eating fresh, freezing, and preserves; strong plants; tolerates wet conditions.
Tillikum	June–Oct.	West	Small to medium, semi-tart fruit with excellent flavor; good for eating fresh and preserves; high yields.
Tribute	June–Oct.	Both	Medium-size fruit with excellent color, firmness, and flavor; excellent for eating fresh and freezing.
Tristar	June–Oct.	Both	Small to medium-size fruit with good color, flavor, and firmness; excellent for eating fresh and freezing; fall crop is heaviest.

Note: Day-neutral cultivars can be grown in many places in an annual production system. However, they will not survive well east of the Cascades in a perennial system.

GOOD STRAWBERRY CULTIVARS FOR THE PACIFIC NORTHWEST

JUNE-BEARING CULTIVARS			
CULTIVAR	**SEASON**	**EAST/WEST**	**CHARACTERISTICS**
Benton	Mid	Both	Medium to large, soft, sweet fruit; excellent for eating fresh and for preserves; poor texture when frozen; very productive.
Hood	Early	Both	Large, firm, bright red, and sweet; good fresh or for preserves; not very winter hardy; very susceptible to viruses; originated in Oregon, where it is leading commercial cultivar.
Puget Reliance	Mid	West	Large fruit; very good for eating fresh; excellent for processing; vigorous, very productive plants; developed in Washington.
Rainier	Late	West	Medium to large, uniformly red fruit, rich flavor; excellent for eating fresh, freezing, and preserving.
Shuksan	Mid	Both	Large, firm, dark red berries with good flavor; excellent for freezing; winter hardy; ideal for British Columbia and northwest Washington.
Totem	Mid	West	Medium to large, solid red berries; a favorite for commercial processing; developed in British Columbia; most widely planted cultivar in the Northwest.

Raspberry

Blackberry

Raspberries, Blackberries, and Other Bramble Berries

*O, blackberry tart, with berries as big as
your thumb, purple and black, and thick with juice,
and a crust to endear them that will go to cream in your
mouth, and both passing down with such a taste that will
make you close your eyes and wish you might live forever
in the wideness of that rich moment.*

—Richard Llewellyn (1907–1983)

Recipes Using Raspberries, Blackberries, and Other Bramble Berries

Raspberries, Blackberries, and Other Bramble Berries

My friend and I head for her back garden raspberry patch below the quince tree. The air is still and hot. With baskets and bowls, we squeeze between the trellises, picking ripe berries hidden among the leaves. Any fruit that doesn't release easily from the central core, we leave to ripen a day or so longer. On the other hand, some berries practically fall off the stem. Of course, these would get squished in the bowl, so they must be popped straight into our mouths. The soft, rich texture and tangy sweet flavor assure us that summer is here. Finally, red-fingered, we head inside with our piles of rubies and dreams of raspberry shortcake.

Flash back to Vancouver Island, British Columbia, and my father-in-law's secret blackberry spot, way up in the hills behind Qualicum Beach. Attired in heavy pants, long sleeves, boots, and sun hats, we follow his lead to the chosen spot. Brambles grow helter-skelter over the logged area and everywhere we find sun-warmed, shiny, black berries with a sweet, winey fragrance I still remember today.

———————

A bramble berry is one of nature's wonders, made up of many small individual drupelets, each with a tiny seed inside. Raspberries, blackberries, and their often thorny kin are members of the *Rubus* genus. This extensive group—referred to as bramble berries or caneberries—Includes blackberry cultivars, such as the Marion and Olallie berries, as well as the raspberry-blackberry hybrids, such as the Tayberry and the Logan and Boysen berries.

To tell the difference between a blackberry and its close relative, the raspberry, pluck a berry from the vine. If the core is still attached to the center of the fruit, it is a blackberry. If the core remains on the vine, leaving the berry with a hollow center, it's a raspberry.

The Pacific Northwest is, by far, the largest commercial raspberry- and blackberry-growing region in the country. But because these berries are so soft and fragile, the bulk of the crop is processed for juice and preserves or flash frozen as individual berries.

Raspberries

Most raspberries available today are complex hybrids of the long, conical berry of the European species *Rubus idaeus idaeus* and the round berry of the American species *Rubus idaeus* var. *strigosus*.

Raspberries are especially well suited to the cool climate west of the Cascade Mountains, somewhat less well to the eastern part of the region. Fields of bright red-fruited Meekers and Willamettes grow throughout the region, particularly in the area from Whatcom County north into British Columbia, and in Oregon's Willamette Valley.

But not all raspberries are red. Black raspberries, known colloquially as "blackcaps," grow extensively in the region, cultivated and in the wild. These small, round bluish black berries, native to North America, have small seeds and a piquant, sweet-tart flavor. Golden raspberries, sometimes called Yellow or Amber, have the same spectrum of flavors as their red cousins. Purple raspberries, a cross between the black and red raspberry, have a subtle, mild flavor.

For a striking presentation, it's hard to beat a dish of fresh black, red, and golden raspberries in a fruit syrup or with a dollop of crème fraîche.

Blackberries

Basically, there are two types of blackberries. One is the trailing, ground-running kind, which tends to take over clear cuts and woodlands, and thrives on the wet west side of the Cascades and northern Sierra range. The other type is the stiff, erect plant with arching canes and good resistance to cold, which does well on the colder, more arid east side of the mountains.

Three wild blackberries are particularly pervasive in the Northwest:

The small native blackberry, *Rubus ursinus*. Known variously as Pacific blackberry, trailing blackberry, Pacific dewberry, or dewberry, it is found in sunny, logged-over forests. The flavor is outstanding, whether plucked and eaten out of hand or baked into a pie or cobbler.

The Himalayan blackberry, *Rubus armeniacus* (formerly *Rubus procerus*). A domestic berry gone wild, it is the most common blackberry in areas disturbed by farming and construction. In 1885, plant geneticist Luther Burbank christened it Himalaya, from where he obtained the seeds; he later discovered they were actually from western Europe. Plants are thorny and extraordinarily vigorous, with large, juicy berries that are pleasantly sweet. The seeds, however, are many and large. This plant is not grown commercially, but savvy pickers gather the fruit in vacant lots, along abandoned train tracks, and in the wild.

The Evergreen or cut-leaf blackberry, *Rubus laciniatus*. Another escapee, easily identified by its lacy, finely cut leaf. It is found, like the Himalayan berry, along the side of the road and in logged-over forests. The glossy black, sweet-tart berry is ideal for pies and other baking.

In 1926, a thornless Evergreen blackberry plant was discovered in Oregon. Except for the lack of thorns, it was otherwise identical to the thorny plant. Growers value its high productivity and ease of picking; however, the harvest is very late, making the weather a potential problem. The berries are firm and sweet, but the seeds tend to be large. There are now many thornless cultivars.

Blackberry × Raspberry Hybrids

A hybrid is the offspring of two plants of different varieties, species, or genera. Sometimes hybrids occur in nature, but plants can also be bred to obtain specific desirable characteristics, such as disease resistance, heavy crop, or rich flavor. Hybrids cannot reproduce themselves.

The following berries are direct hybrids of one or more raspberries and one or more blackberries; that is, one parent of the cultivar is a blackberry and the other is a raspberry. All are grown, or have been grown, commercially in the Pacific Northwest.

Boysen. Probably the result of blackberry, raspberry, and Logan crosses made by Rudolph Boysen in Napa, California, in the 1920s. Grows like typical trailing blackberries. Large, reddish-purple berry, with big drupelets. Succulent, especially when cooked. Widely grown in Oregon.

Logan. The oldest trailing caneberry cultivar on the Pacific Coast. Thought to be a hybrid between the Red Antwerp red raspberry and a trailing blackberry, developed by backyard plant breeder Judge Logan of Santa Cruz, California, in the 1880s. A productive plant with large, elongated, deep red or maroon berries with a unique, tart flavor. Now grown primarily in Oregon.

Tayberry. A cross, in 1977, between a raspberry and Aurora, a trailing blackberry, near the Tay River in Scotland. Similar to the Logan. Long, firm, purplish red berry with a vibrant woodsy flavor.

Bramble Berry Picking

Is it possible to have too many bramble berries?

Sweet, ripe raspberries are delicious right off the vine. Pick them when fully ripe and handle gently. Raspberries, more than any other berries, require a shallow picking container, so as to not crush the fruit.

Blackberries are ready to pick when they turn from shiny purple-red to a dull black. Pop one in your mouth, just to make sure. Pick only the ripe ones, saving the "almost readys" for next time. Then hurry your fragile treasures into the kitchen and refrigerate them to enjoy at your leisure.

Growing Bramble Berries in the Home Garden

Raspberries and blackberries are two of the easiest berries to grow. By the third year after planting, you will probably have a sizeable crop, which then will continue season after season for many years.

The long, mild growing season in the Pacific Northwest is ideal for all types of bramble berries. Numerous kinds thrive in the relatively cool, moist summers and mild winters found on the west side of the Cascade mountains. East of the Cascades, where summers are hotter and winters are colder, gardeners also have a choice of cultivars.

Bramble berries do best with full sun, good air circulation, and a rich, loose soil that drains well. Trellises are essential to corral raspberries and trailing blackberry vines.

Buy certified disease-free plants from a reputable nursery. Plants from a neighbor may carry a virus or root rot.

All raspberries and most blackberries are self-fertile, that is, additional cultivars are not required for cross-pollination. However, by selecting several different cultivars that ripen at various times, you'll have a wider selection of fruit and a longer harvest.

Two types of raspberries are available to the home gardener:

The summer-bearing cultivars. These produce fruit only on year-old canes, yielding one good-size crop a year, usually in July.

The ever-bearing cultivars. These are typically grown for a late summer–early fall crop. However, if the canes are cut back only as far as they fruited in the fall—but not to the ground—they will also produce a summer crop.

Black raspberries can be successfully grown in the Pacific Northwest. However, you will find black raspberries are generally shorter lived than red raspberries and much shorter lived than blackberries, due to disease susceptibility.

Blackberries are a little less fussy about soil conditions than raspberries, and once established, they'll produce crops for fifteen to twenty years. Three types of blackberries are available to the home gardener:

The trailing blackberries, also called dewberries. Trailing blackberries will ramble all over unless tied to a fence or trellis. They include various hybrids, such as Boysen, Logan, and Tayberry, and for easy picking, the Thornless Evergreen.

Semi-erect plants. These plants also must be trellised. Semi-erect blackberries include Chester Thornless, which is large, sweet, and very productive, among others.

Erect plants. Fairly easy to control, erect blackberry plants include Arapaho, which is thornless and early ripening, and Navaho, also thornless but late ripening.

GOOD BRAMBLE BERRY CULTIVARS FOR THE PACIFIC NORTHWEST

RASPBERRY CULTIVARS: SUMMER BEARING (*FLORICANES*)			
CULTIVAR	**SEASON**	**EAST/WEST**	**CHARACTERISTICS**
Red Raspberries			
Canby	Early	West	Large, light red fruit with excellent flavor; very few spines; developed in Oregon.
Chilliwack	Mid	West	Extra large, red fruit with excellent flavor; resists root rot; more tolerant of wet sites and poor drainage than most raspberry cultivars; developed in British Columbia.
Meeker	Mid	West	Large, soft and juicy, dark red berry with excellent sweet flavor; protect plant in coldest areas; developed in Washington.
Tulameen	Late	West	Very large, red fruit with excellent flavor and firmness; good yield, attractive plant; developed in British Columbia.
Willamette	Early	West	Large, dark red fruit, soft, juicy texture with tart flavor; high yield, vigorous plants with many spines; needs staking; avoid dry areas east of the Cascades; developed in Oregon.

GOOD BRAMBLE BERRY CULTIVARS FOR THE PACIFIC NORTHWEST

CULTIVAR	SEASON	EAST/WEST	CHARACTERISTICS
RASPBERRY CULTIVARS: FALL BEARING "EVERBEARING" (*PRIMOCANES*)			
Red Raspberries			
Autumn Bliss	Early	Both	Very large, tasty, red fruit; less prone to mildew; generally very cold-tolerant; developed in England.
Heritage	Late	Both	Medium-size, mild flavor, firm, red fruit; produces berries over wide climatic zones and soil conditions; developed in New York.
Summit	Early	Both	Small, flavorful, red fruit; productive; resistant to root rot; will tolerate wet sites; developed in Oregon.
Golden Raspberries			
Fallgold	Early	Both	Medium to large, golden fruit with a mild, sweet flavor; vigorous; good yield, bears in summer and into fall; developed in New Hampshire.
Black Raspberries			
Black Hawk	Early	Both	Large, black fruit; very spiny, susceptible to anthracnose; grow far from other raspberries; developed in Iowa.
Cumberland	Early	Both	Large, firm, black fruit with very good flavor; vigorous plant; resists bushy dwarf; suffers from virus diseases; developed in Pennsylvania.
Munger	Mid	Both	Medium-size, shiny black fruit, sweet flavor; the most widely grown black raspberry in the Pacific Northwest; developed in Ohio in 1897.

GOOD BRAMBLE BERRY CULTIVARS FOR THE PACIFIC NORTHWEST

BLACKBERRY CULTIVARS			
CULTIVAR	**SEASON**	**EAST/WEST**	**CHARACTERISTICS**
Arapaho	Early	Both	Medium-size, firm fruit with excellent flavor and small seeds; thornless; moderately vigorous erect plant; hardy to −10° F.
Black Butte	Early	West	Very large, attractive black fruit with good flavor, berries are twice the size of Marion; trailing canes produce good yields; a new cultivar; developed in Oregon.
Black Douglass	Early	West	Large, long black fruit with Marion-like flavor; excellent for eating fresh and for preserves; thornless, vigorous, trailing canes; developed in Oregon by amateur breeder Barney Douglass.
Cherokee	Early	Both	Large, sturdy, firm, glossy black fruit with excellent flavor; similar to Marion but firmer; erect thorny cultivar, vigorous; hardy to −10° F.
Chester Thornless	Late	Both	Very large, firm, sweet fruit; thornless, semi-erect, productive plant; bears over a five- to six-week season; hardy to −10° F.
Kotata	Mid	West	Large, black, fairly firm with excellent flavor; similar to Marion; high yield, very thorny, trailing plant; developed in Oregon.
Marion	Mid	West	Large, shiny dark pink to black, soft fruit with excellent tangy-sweet flavor and less noticeable seeds; the standard for comparison; productive, thorny, trailing plant; from a cross between two blackberries, Olallie and Chehalem, developed in Oregon and named for Marion County in 1956.

GOOD BRAMBLE BERRY CULTIVARS FOR THE PACIFIC NORTHWEST

BLACKBERRY CULTIVARS (cont'd)			
CULTIVAR	**SEASON**	**EAST/WEST**	**CHARACTERISTICS**
Navaho	Late	West	Small, firm fruit with excellent flavor; thornless, erect; hardy to 0° F.; developed in Arkansas.
Olallie	Early	West	Long, slender, sweet-tart berries with excellent flavor, glossy black when ripe; trailing plant developed in Oregon from a Young and Black Logan cross, now grown mostly in California.
Thornless Evergreen	Late	West	Medium, black, firm fruit, over 1 inch long, very sweet, with large seeds; pair with a more flavorful berry for pies; vigorous, drought-resistant trailing plant; hardy to 8° F.
RASPBERRY–BLACKBERRY HYBRIDS			
CULTIVAR	**SEASON**	**EAST/WEST**	**CHARACTERISTICS**
Boysen	Mid	West	Very large, reddish black, soft fruit with a sweet-tart flavor, fruity aroma, and large seeds; ideal for eating fresh or for freezing; trailing type, thorny, vigorous canes; also thornless variety; fruit ripens over several months; hardy to about 10° F.; developed in California.
Logan	Early	Both	Large, elongated, dull-colored fruit, highly acidic; trailing type; thornless is less productive than Marion; winter protection needed for eastern Washington; developed in California.
Tayberry	Mid	West	Large, reddish purple, elongated, tart-flavored fruit; especially good when cooked; extremely thorny, needs winter protection in colder areas; developed in Scotland.

Blueberry

Huckleberry

Blueberries, Huckleberries, and Crowberries

Blueberry pie by the Fourth of July.

—An Old Alaskan Saying

Recipes Using Blueberries, Huckleberries, and Crowberries

Blueberries, Huckleberries, and Crowberries

One of my sisters has blueberry bushes in her backyard, set between the sunflowers, tomatoes, and lemon cucumbers near the house, and the row of fruit trees near the back fence. The blueberries are about shoulder-high and protected from the birds by sturdy, moveable cages. On summer mornings, she darts out before breakfast, often leaving footprints in the dew, to gather a handful of dusky blueberries, a few raspberries hanging over the fence, and if she's lucky, a tree-ripened peach.

Another sister. Another time.

It was my first trip to Alaska. My sister, her kids (all grown now), and I head out for some berry picking. She drives to a field of tall grasses, with caribou grazing in the distance, and the aptly named cotton flowers bobbing on the edge of the meadow. The shiny purple-black crowberries grow low on the marshy ground and are pretty easy to pick. Frankly, they don't taste like much when fresh. But that's not the point. These berries are going into my sister's famous crowberry pie. And once cooked, the rather insipid raw crowberries would be transformed into tasty morsels.

When it came time to say good-bye to the land of the midnight sun and my sister's good cooking, she mentioned that either blueberries or huckleberries would work just fine for her pie, knowing, of course, that I was soon to leave crowberry country.

Blueberries are indigenous to North America, where they have been a part of the food heritage since long before the first settlers arrived from across the Atlantic.

The early colonists, particularly those in New England, enjoyed the wild blueberries, which were similar to their familiar English bilberries, and baked them into slumps and grunts, buckles and betties.

The native peoples taught the newcomers how to dry the fruit for winter use. In the early 1800s, when Lewis and Clark arrived in what was later called the Pacific Northwest, they found the inhabitants preserving the wild berries by smoke-drying them. Even today tiny huckleberries are harvested and used as ceremonial food by the local native Americans.

Blueberries were also said to have medicinal value. For example, the early settlers would administer a syrup of blueberry juice to cure a cough. This remedy may well have worked, since, as we now know, blueberries are a good source of vitamin C and vitamin A.

There is a lot of confusion about what is a blueberry, and what is a huckleberry.

In common usage, the lighter colored, large berries are called blueberries and the smaller, darker, and more acid-tasting ones are called huckleberries. Generally, the two can be used interchangeably. The real way to find out what you have in hand is to taste one.

Both blueberries and huckleberries are members of the heath family, *Ericaceae*. These berries, of the *Vaccinium* genus, have tiny, unnoticeable seeds, while huckleberries, of the *Gaylussacia* genus, not generally found here in the Northwest, have ten hard, seedlike nutlets inside.

But this is not as easy as it seems. For example, the tart wild red huckleberry, despite its common name, is related to the blueberry, not to the true huckleberry of the southeastern United States.

The crowberry, *Empetrum nigrum*, is related to neither the blueberry nor the huckleberry. An evergreen, creeping, heathlike shrub, the crowberry is found throughout Alaska and northern British Columbia. The berry is small, round, and shiny black. It is edible raw, but the flavor improves with cooking.

Blueberries

Until the early part of the twentieth century, all blueberries were wild. Then, in 1910, Frederick V. Coville, a botanist with the USDA (U.S. Department of Agriculture), embarked on what came to be the transformation of a simple wild fruit into a commercially viable, cultivated plant. He and Elizabeth White, a New Jersey cranberry grower who provided the land for the project, selected plants with such favorable characteristics as resistance to diseases, cold hardiness, good berry size, and, naturally, flavor. For the next twenty-seven years, Coville hybridized new varieties and encouraged their development. The Coville berry, a top commercial blueberry, is still available to home gardeners.

Blueberries generally are categorized as highbush or lowbush. The large, dusky blue, fresh blueberries found in the supermarket produce section are, most likely, highbush blueberries. Those found in cans are probably lowbush. Breeders have also crossed lowbush and highbush species, to what are called half-highs, by combining the large berry of the highbush parent and the wild flavor of the lowbush. These half-highs are grown only in areas that have a very cold winter. There are others that grow in specific areas. Types of blueberries include:

The highbush blueberry, *Vaccinium corymbosum*. Native to the Atlantic coastal plain, these plants can reach up to ten feet or more. This is the most important commercially grown blueberry species in the country and an ideal berry for the Northwest climate. Cultivated since the early part of the century, highbush berries make up well over two thirds of the North American blueberry crop. Many are picked by hand for the fresh market. The majority, destined for the processing plant, are harvested by machine.

The lowbush blueberry, *Vaccinium angustifolium* (and to some extent *Vaccinium myrtilloides*). Also native to the northeastern United States, primarily Maine and the adjacent parts of Canada, these plants grow only a foot or two tall. The berries are small and dark. They make up a third of the country's total blueberry crop. The lowbush plant is a semi-wild plant that spreads by underground stems. Pickers harvest most of the shiny, flavorful fruits with a blueberry rake or a machine that works only on flat, nonrocky ground. The crop, mostly from Maine, is almost entirely for processing as canned fruit and for breakfast cereal and muffin and pancake mixes.

The rabbiteye, *Vaccinium ashei*. This type of blueberry grows well in the southeastern United States. It is known for its heat and drought tolerance but is less resistant to cold than other cultivars.

The red huckleberry, *Vaccinium parvifolium*. Related to the common blueberry and with similar foliage, this berry is found growing wild along the north Pacific coast and particularly in Alaska. This berry is bright red, just one-quarter inch across, quite sour, but excellent for making jams and jelly.

Huckleberries

Several kinds of wild huckleberries grow in the Northwest. All are in the heath family, *Ericaceae*. These are not the small, dark huckleberries of the *Gaylussacia* genus found in the southeastern United States, with hard, seedlike nutlets inside. Western huckleberries have tiny, unnoticeable seeds. Among them are:

The blue huckleberry, *Vaccinium membranaceum*. The dominant huckleberry in the Cascades, this is an erect, deciduous shrub found throughout the area, including the Indian Heaven Wilderness and the Mount Adams huckleberry fields north of the Columbia River.

The evergreen or California huckleberry, *Vaccinium ovatum*. A compact shrub with dark lustrous leaves and tasty black berries good for baking.

The red huckleberry, *Vaccinium parvifolium*. Though called a huckleberry, this plant is actually a blueberry (see above).

Crowberries

The crowberry, *Empetrum nigrum*, is unrelated to the blueberry. This low-growing shrub, found primarily in Alaska, has tiny black berries that are at their best when cooked or mixed with other berries.

Picking Blueberries and Huckleberries in the Wild

Berries ripen first at lower elevations, and then progressively later, the higher you hike. The most flavorful fruit will be fully colored and release easily from the stem.

Picking berries in the wild seems simple, but there are some hazards.

• Some national forests and national parks require a permit. Check first.

• Some berries are toxic. Make sure that the berry you want to pick is edible. Get an expert's opinion before you taste.

• Huckleberries and blueberries are a favorite food of bears. Give all wildlife plenty of room. Some pickers sing or wear bells to announce their presence.

Growing Blueberries in the Home Garden

Blueberries are both an easy-care, ornamental shrub and a delectable fruit. The yearly cycle starts in the spring with small white or pinkish urn-shape flowers. Summer brings decorative and edible fruit. Then, as the show continues into autumn, there is a display of red or amber foliage before winter arrives.

To deliver all this bounty, blueberry plants require an acid soil high in organic matter, plenty of water, lots of sun, and a plan to deal with the birds. Diseases are uncommon in the home garden.

Many blueberry cultivars are considered self-fertile. However, since many are not, you can ensure a good fruit set and large berries by planting more than one cultivar. To extend the harvest, choose a selection of early, midseason, and late cultivars.

Plant blueberries in the early spring. It will take four or five years before young plants begin to bear fruit. But once established, a single berry bush can produce twenty pounds of fruit during the season.

Blueberries are well suited to the climate on the west side of the Cascades. East of the mountains, they will grow in the upland areas, but the yield per plant will not match the berries grown in the west.

The harvest starts in June in Oregon and continues through mid-September, depending on the cultivar. Blueberries grow in clusters and ripen over several weeks. The fruit, however, may turn a beautiful blue well before it is fully ripe. Taste before you start wholesale picking, to make sure the berries are sweet and fully mature. When they are ripe, you can just tickle the fruit off the plant right into your palm.

GOOD BLUEBERRY CULTIVARS FOR THE PACIFIC NORTHWEST

CULTIVAR	SEASON	CHARACTERISTICS
Bluecrop	Mid	Medium-large, firm, light blue, flavorful fruit; great for eating fresh; productive, vigorous plant that bears fruit for a month.
Chandler	Late	Extremely large berry; light blue, high quality; very productive over a 4- to 5-week period.
Coville	Late	Medium blue, very large, aromatic berries with tart flavor; berries do not drop from bush; vigorous and productive plant.
Darrow	Late	Light blue, tart, extremely large berries of excellent quality; fruit ripens in August; vigorous, upright bush.
Duke	Early	Excellent yield and fruit quality; mild flavor; stores well; one of first to ripen.
Earliblue	Early	Excellent quality, large fruit; first to ripen; upright bush.
Elliott	Very Late	Powder blue, medium-size, mild-flavored, firm berries; ripens late August, if enough heat; produces until first frost.
Jersey	Mid	Small fruit, excellent flavor; processes well, good for canning.
Spartan	Early	Very large, light blue, firm berries; best flavor of early cultivars; heavy bearer.
Toro	Mid	Large fruit; outstanding quality; very productive, stocky plant; grows well on both sides of the Cascades.

Currant

Gooseberry

Currants, Gooseberries, and Jostaberries

Gently cook the gooseberries, rub them
through a fine sieve, and add the pulp to the butter sauce.
This sauce is excellent with grilled mackerel and the
poached fillets of that fish.

—Auguste Escoffier (1846–1935)

Recipes Using Currants,
Gooseberries, and Jostaberries

Currants, Gooseberries, and Jostaberries

My real introduction to gooseberries occurred on Whidbey Island while visiting friends. It was a warm summer day with a view of Mount Baker across Puget Sound. Our friends suggested we come see the back garden, especially the gooseberries along the fence, which were supposedly ready to pick.

What a surprise to find that the funny name *Stachelbeeren* I learned in high school German for these lovely, translucent green berries, translates as "sticker berries." And stickers they were. (I have since learned that not all gooseberries have thorns.)

Red currants, on the other hand, always make me think of the gaiety of Vienna and the indispensable Austrian *Ribisel*, which is the foundation of the rich currant sauce served with venison and other game and which festively gilds the fancy tortes, all sugar and cream, to be served, of course, with a mid-afternoon coffee.

Hiking in Austria, I often saw red currants, easily identified by their distinctive three-lobed leaf. The first time I saw one growing, I knew that it was a red currant and that it was edible, but I had no idea how astringent the raw fruit could be.

———

Both currants and gooseberries are of the *Ribes* genus. Unlike in Britain, where these fruits are taken seriously, finding currants and gooseberries in an American market may take some searching. Try the specialty produce markets and farmers' markets in mid-summer for crisp green or burgundy red gooseberries with smooth or fuzzy skins, jewel-like red and white currants, and the distinctively aromatic black currants. Most of the country's commercially grown red currants are centered in the Northwest. The harvest is short, however, just a few weeks, from early July to early August.

Gooseberry season runs a bit longer, generally from June through August, plus the occasional import from New Zealand in the late fall and into December. A few commercial growers do raise gooseberries in British Columbia, Oregon, Washington, and northern California. In Oregon, however, many berries are destined to be canned (for year-round use), rather than for the fresh market.

Currants

When the English colonists arrived in Massachusetts, one food they sorely missed was the red currant. They repeatedly requested that plants be shipped, but when the plants finally arrived they did poorly. The English cultivars were, perhaps, unsuited to the American climate.

Red, white, and black currants thrive in the cool, moist northern regions of the world, such as Great Britain, Ireland, Scandinavia, Germany, France, and luckily for us, the Pacific Northwest.

There are three types of currants:

Red currants, *Ribes rubrum* and *Ribes sativum*. Like shiny rubies, these are the most common of the three. Toss them into a green salad or add them to a mixed berry compote for a bit of bite. When they are fully ripe, enjoy them *au naturel* or doused with cream and a liberal sprinkling of sugar.

White currants, *Ribes rubrum* and *Ribes sativum*. These include various shades of cream, off-white, and a buttery yellow. They are essentially red currants without the red color. The berries are translucent and slightly sweeter than their red counterparts. Red and white currants can be used interchangeably in recipes.

Black currants, *Ribes nigrum* and *Ribes odoratum*. These are different species altogether from the red. The fruit is dark purple, almost black, with a distinct fragrance and robust flavor. Generally black currants are considered too bitter to eat fresh. In Europe black currants are widely grown for their juice, much of which is turned into black currant syrup or *crème de cassis*, a liqueur to pair with white wine for kir, or with sparkling wine for a kir royal.

Medicinal properties of black currant berries and leaves were touted in the fourteenth century. The claim that the berry could soothe sore throats and other coldlike symptoms predated the discovery of vitamin C.

Note that two different fruits go by the name currant. The berries of the black currant plant are *not* related to the tiny dried grapes sold as black or dried or Zante currants. Their name is derived from Corinth, the Greek port from where the grapes were originally shipped.

Gooseberries

Gooseberries show up everywhere on the menu. The French serve these berries with mackerel. That dish is called *groseille à maquereau*, which, curiously, is also the French name for gooseberry. The name in English suggests that this berry, at one time or another, must have accompanied goose. The most famous gooseberry dish, however, may be the inspired pairing of lightly sweetened berries with silky-rich whipped cream, the classic gooseberry fool.

There is just one species of gooseberry, *Ribes grossularia*, though there are many cultivars. Translucent, green, gold, pink, or purple, round or oval, the size, color, and shape of gooseberries depend on the cultivar. Rich in natural pectin, gooseberries transform meat juices into savory sauces and thicken up preserves and pie fillings.

Just a few gooseberry cultivars, Poorman for one, are sweet enough to eat out-of-hand. Be warned that most are puckeringly tart.

Jostaberries

The jostaberry (pronounced YOH-sta berry) is the result of complicated crosses between the black currant and the gooseberry, taking advantage of the best qualities of each. The breakthrough for the Jostaberry came in Germany in the late 1950s from the work of Dr. Rudolph Bauer. The name came from the first letters of two German words: *Johannisbeeren* (currants) and *Stachelbeeren* (gooseberries). There is only one species:

Jostaberry, *Ribes nidigrolaria*. Less subject to disease than either currants or gooseberries. Jostaberry shrubs are totally thornless. The fruit is formed in clusters. At first a Jostaberry looks like a gooseberry, then as it ripens, it turns a purplish black. It is two or three times the size of a red currant, but smaller than a large gooseberry. The flavor is less sharp than its black currant parent.

There is now also a red Jostaberry, a hybrid of the red currant and the gooseberry.

Picking Gooseberries

Gooseberries must be left on the bush to turn sweet. Some green varieties turn amber as they ripen. For jam, pick the berries a little less ripe than for eating fresh. Choose firm, dry berries. They will keep in the refrigerator crisper for at least a week. To prepare the berries for cooking, "top and tail" them, as the English say, meaning to snip off the stems and blossom ends, usually with scissors.

Growing Currants and Gooseberries in the Home Garden

One way to ensure a supply of fresh currants and gooseberries is to plant your own. They are relatively easy to grow if you select disease-resistant cultivars.

Currants, gooseberries, and Jostaberries are well suited to the cool, moist climate of the Pacific Northwest. All three make attractive ornamental shrubs, and there are now new cultivars of gooseberries with fewer thorns and larger berries that are easier to pick. And some of these cultivars are resistant, or nearly resistant, to powdery mildew. The plants come into their own in summer when small, tart berries cover the bushes, leading to thoughts of jams and jellies, preserves and pies.

Currants and gooseberries are hardy down to about minus thirty degrees Fahrenheit or so, but they bloom very early and are, therefore, susceptible to injury from late spring frosts. The fruit is susceptible to sun scald, especially during sudden hot spells.

The plants thrive in the cool, moist coastal areas from British Columbia to the northern coast of California. These berries also do well in protected areas east of the Cascades.

In the early twentieth century, white pine blister rust was introduced to North America, and *Ribes* berries, both wild and cultivated, were the alternate hosts of this fungal disease. In an attempt to save the valuable white pine forests, growing *Ribes* berries was banned in most areas of the United States. The ban is now lifted in most areas, including the Northwest. Several black currant cultivars, such as Consort and Crusader, are available in rust-resistant hybrids.

Two other problems plague the currant plants: imported currant worm, which appear in large numbers and can devour the leaves on a bush in a few days, and currant fruit fly, which lays eggs that mature inside the ripening berries, making them unfit to eat. Other than these problems, gooseberries and currants are among the easiest berries to grow.

GOOD CURRANT AND GOOSEBERRY CULTIVARS FOR THE PACIFIC NORTHWEST

CURRANT CULTIVARS		
CULTIVAR	**SEASON**	**CHARACTERISTICS**
Red Currants		
Jonkheer van Tets	Early	Large, very flavorful fruit; heavy producer; excellent for sauces and jelly; resistant to powdery mildew; originated in the Netherlands.
Perfection	Early	Extra large, bright crimson fruit, sweet flavor; resistant to white pine blister rust and powdery mildew; productive.
Red Lake	Mid	Medium to large, light red berries; easy to pick; productive; popular choice.
Wilder	Late	Small, dark red, low-acid berries in large, compact clusters; fruits over a long season; susceptible to powdery mildew.
White Currants		
White Imperial	Early	Medium, translucent white to creamy yellow berries, very sweet and juicy; good for eating fresh; resistant to powdery mildew.
Black Currants		
Ben Sarek	Early	Large fruit, compact shrub with high yield; resistant to powdery mildew; frost resistant.
Consort	Early to Mid	Small to medium, blackish purple berry with distinctive flavor; resistant to white pine blister rust, but very susceptible to powdery mildew.
Crusader	Mid to Late	Small to medium black berries with fair flavor; resistant to white pine blister rust; requires cross-pollination; developed in Ottawa, Canada, about 1950.
Note: Also recommended are Ben Lomand and Ben Connan, two more of the newer Scottish black currant cultivars, generally resistant to powdery mildew and leaf spot.		

GOOD CURRANT AND GOOSEBERRY CULTIVARS FOR THE PACIFIC NORTHWEST

GOOSEBERRY CULTIVARS		
CULTIVAR	**SEASON**	**CHARACTERISTICS**
Captivator	Early to Mid	Small, dull red berries, moderately sweet, good flavor; practically thornless; resists powdery mildew; very winter hardy; developed in Ottawa, Canada.
Hinnonmaki Red	Mid	Medium, burgundy red or purple fruit, good flavor; thorny; resistant to powdery mildew; productive; originated in Finland.
Oregon Champion	Early	Medium, pale green fruit, good flavor; heavy yielding, thorny bush, 3 to 5 feet; resistant to white pine blister rust; originated in Salem, Oregon, circa 1860; widely grown in the Pacific Northwest.
Pixwell	Mid	Small to medium pink-green berries with thin skin, very juicy; excellent for pies and preserves; few thorns, easy to pick (as name indicates) since berries hang below leaves; resists powdery mildew and white pine blister rust; very productive; extremely winter hardy; developed in Fargo, North Dakota, from Oregon Champion, introduced in 1932.
Poorman	Late	Medium to large tear-shaped red fruit, sweet (one of the sweetest) with very good flavor; eat fresh or cooked; resists powdery mildew and white pine blister rust; vigorous and productive; favorite for snowy areas of the west; originated in Utah.

Lingonberry

Cranberry

Cranberries and Lingonberries

*I said my prayers and ate some
cranberry tart for breakfast.*

—William Byrd of Westover, Virginia (November 11, 1711)

Recipes Using Cranberries and Lingonberries

Cranberries and Lingonberries

My first visit to a cranberry bog was many summers ago in Massachusetts at the Cape Cod National Seashore. I discovered wild plants nestled among the sand dunes and later took a Park Service trail out to see a cultivated bog. It was too early in the season to see the ripe berries, but I did find the last of the tiny pale-pink blossoms that supposedly gave the cranberry its name. Legend has it that the early German and Dutch settlers in the area called the fruit *Kraanbere* (craneberry) because the shooting star-like cranberry blossom resembles the shape of a crane's head. Eventually, the name became cranberry.

More recently I visited the coastal town of Grayland, Washington, for some biking and a close look at the cranberry bogs. These creeping, low-growing cranberries, and those up the coast in North Beach, are the cream of the crop. They are dry-picked and destined for the fresh berry market, while the majority of the nation's crop is wet-harvested and suitable only for processed foods.

The lingonberry seems like the uninhibited little sister of the commercially grown cranberry. Smaller, with a spicy, wild-tart flavor, these berries are hard to come by in the United States, unless you count the lingonberry sauce on the grocer's shelf. Or live in Alaska.

A few years ago, my husband and I were on a wildflower walk in Denali National Park, within view of the cloud-shrouded Mount McKinley. On the steep hillside, moist with melting snow, the flowers were everywhere, as well as a good number of edible berries, including the teeny, red, jelly bean-like lingonberries.

My sister, who lives in Alaska, keeps a stash of lingonberries in her freezer. Then, at pancake time, she tosses a handful of the still-frozen lingonberries into the batter. "Better than cranberries," she boasts. "They don't need chopping."

Alaska, of course, isn't the only place that lingonberries grow. They are highly prized throughout northern Europe, especially in the Scandinavian countries and Germany. It was the Swedes who bestowed the name.

Cranberries and lingonberries are among the few indigenous fruits of North America. Across the northern tier of the continent, from Massachusetts to Wisconsin and north into Canada, the native peoples treasured the cranberry, known variously as *sassamanesh*, *ibimi* (meaning bitter berry), and *atoqua*.

Long before Columbus arrived, they boiled cranberries with maple syrup to make them more palatable and dried them for winter use. They also made pemmican, an early version of energy bars, by blending dried venison and fat with crushed cranberries, pounding the mixture into a paste, and then preserving it in the form of small pressed cakes.

Fresh cranberries are also known for their excellent keeping qualities. Way before modern refrigeration, sailors packed cranberries in barrels of water and shipped them to Europe. Scientists now know that the long-keeping cranberry contains large amounts of benzoic acid, a natural preservative. This transatlantic trade in cranberries may have been the first American export. In addition, on long voyages the crew relied on cranberries to prevent scurvy, much as the British sailors ate limes, long before the discovery of vitamin C.

The cranberry was one of many native foods served at the first Thanksgiving, a three-day celebration of thanks that took place at Plymouth, Massachusetts, in October, 1621. The banquet included venison and wild turkey, pumpkins and Indian corn, and blueberries and cranberries, but no sugar.

During the Civil War, cranberries appeared again at another Thanksgiving. In 1864, just a year after Lincoln proclaimed the first official Thanksgiving, General Ulysses S. Grant ordered a large shipment of cranberries, so his tired soldiers could celebrate a traditional Thanksgiving.

Cranberries

There are basically two types of cranberry:

The American cranberry, *Vaccinium macrocarpon*. This is the cranberry of commerce, red marble-size cranberries piled in bags in the produce section just before Thanksgiving. Native to North America and found in boggy or marshy areas in the temperate north, these are large, flavorful berries, twice the size of the European cranberry.

The European or bog cranberry, *Vaccinium oxycoccus*. Sometimes called the small cranberry, this plant fruits in the late fall. It is a little berry on a scrubby vine that grows in the wild, but it tastes much like the big American cranberry and may be used in the same way.

The cultivation of wild cranberries began about 1816 when a Cape Cod resident discovered that they grew better when the beach sand blew over the bogs. So he fenced in the plants and added more sand to the bogs.

A cranberry bog is not, as is often assumed, full of water, but a layered bed of peat and sand where the vines grow. These beds are flooded only for the harvest and, in the Northeast, during winter for protection against cold damage. Once established, the plants can survive for many decades, occasionally for a century or more. Many cranberry beds planted at the turn of the century, for example, are still in commercial production.

Charles McFarlin started cranberry cultivation in Oregon in 1885, with vines shipped out from Massachusetts. Now most of the Oregon harvest comes from around Bandon, using the deep red McFarlin berry. The new plantings are mostly Stevens.

In the Northwest the harvest starts at the end of September. For the dry harvest, growers use mechanical Furford Pickers, developed in Washington, that look like giant lawn mowers. They comb the berries from the vines and prune at the same time. The berries are quickly transported to Markham, Washington, sorted, and packed in plastic bags. These are the fresh berries sold in the produce section.

For more than a century, fresh cranberries have been sorted by bounce. Only the firm, highest quality berries will bounce over an angled slat; these are destined to go to the fresh market. The also-rans drop vertically into the second-class section for processing.

The majority of the Northwest crop is wet-harvested, slated for use in processed foods. The bogs are flooded with a foot of water, and a mechanical picker, called an egg-beater, paddles along, loosening the cranberries from the vines. As the berries float to the surface, the bog turns into a sea of scarlet. The berries are corralled with wooden booms and trucked to the processing plant.

Ocean Spray, a grower-owned marketing cooperative, started in 1930. It handles about 80 percent of the cranberries sold worldwide. In the Northwest, there are both Ocean Spray and independent growers.

Lingonberries

Whether you call them lingonberries, low-bush cranberries, foxberries, cowberries, or partridge berries, these wild little jewels taste best if picked after the first frost. There are two kinds:

The lingonberry, *Vaccinium vitis-idaea*. A low, creeping bush, producing a tiny, red, oval fruit, a little larger than a pea. The flavor is similar to the American cranberry but not as sharp.

The dwarf lingonberry, *Vaccinium vitis-idaea minus*. The plant grows just a few inches high, with bright pink blossoms and leaves less than a quarter inch long. It is similar to the lingonberry, only smaller.

Growing Cranberries and Lingonberries in the Home Garden

Cranberries and lingonberries are in the same *Vaccinium* genus as blueberries and huckleberries. All require well-drained, acid soil, rich in organic matter.

The American cranberry, *Vaccinium macrocarpon*, produces clusters of white, urn-shape flowers, which then turn into bright red, tart berries, come August or

September. The shrubs grow only about a foot tall, but their capacity to spread is unlimited. Be warned, the sprawling stems will take root wherever they touch the ground, and then grow outward. The berries tend to be bitter before the first frost, then the flavor improves. Cranberries are hearty enough to grow in most areas of the Pacific Northwest.

Plant lingonberries in a shady spot. Lingonberries are partially self-fertile, so two or more cultivars will increase the yield. In cold or dry areas, mulch with pine needles. Water regularly the first year, then back off and give just what is needed.

The standard lingonberry, *Vaccinium vitis-idaea*, grows to about a foot or so tall with heavy crops of pea-size red fruit. The flavor is similar to the true cranberry, but less sharp. Koralle, one example of this type, is popular in Europe. It grows tall and bushy.

The dwarf lingonberry, *Vaccinium vitis-idaea minus*, is a small evergreen ground cover with showy heatherlike flowers twice a year. The red fruit is ready to pick in July and again in late October, but it is pretty sparse. Harvest lingonberries when the fruit is entirely red.

GOOD CRANBERRY AND LINGONBERRY CULTIVARS FOR THE PACIFIC NORTHWEST

AMERICAN CRANBERRY: *VACCINIUM MACROCARPON*		
Very tart, hard, thumbnail-size fruit; bright red when fully ripe; keeps several months at 40 to 45° F.; vines grow 4 to 5 feet tall.		
CULTIVAR	**SEASON**	**CHARACTERISTICS**
McFarlin	Autumn	Medium-size red fruit; an old commercial variety selected from the wild.
Stevens	Autumn	Large red fruit; very productive; a hybrid selection that does well in the Pacific Northwest.
LINGONBERRY: *VACCINIUM VITIS-IDAEA*		
Bright red, cranberry-like fruit but less tart; excellent for preserves, sauces, and syrups; pick when thoroughly red, preferably after the first frost for best flavor.		
CULTIVAR	**SEASON**	**CHARACTERISTICS**
Koralle	July and October	Many red, pea-size fruits; upright and bushy shrubs; usually bears the first year; a European variety.
Red Pearl (European Red)	July and October	Large 1/3- to 1/2-inch, tasty fruit; heavy producer; upright plant, easy to grow; from wild European lingonberries; good pollinator for Koralle.

the
recipes

Breakfast and Breads

Bob's Blueberry Blue Cornmeal–Buttermilk Pancakes

Makes about twenty 3¹/2-inch pancakes

The idea for this recipe came while breakfasting at the Old Historic Taos Inn in Taos, New Mexico, some years ago. It has become a Saturday morning tradition ever since. During the summer, my husband, Bob, adds a few small blueberries to each pancake just after pouring the batter. We like to top the pancakes with maple syrup and warm homemade applesauce.

*3/4 cup buttermilk
1 large egg
2 tablespoons vegetable oil
1/2 cup and 2 tablespoons all-purpose flour
1/2 cup and 2 tablespoons finely ground blue cornmeal
1 tablespoon sugar
2 teaspoons baking powder
1/2 teaspoon salt
1 cup blueberries or huckleberries, rinsed,
  picked over, and blotted dry
Maple syrup and applesauce (optional)*

Whisk together the buttermilk, egg, and oil in a large bowl. Sift or stir together the flour, cornmeal, sugar, baking powder, and salt. Add to the bowl and whisk until smooth.

Heat a griddle or cast-iron comal over medium-high heat. Oil lightly. Spoon the batter, about 2 tablespoons per pancake, onto the griddle and add a few berries to each. When the bubbles have popped, turn, flatten slightly with a spatula, and continue to cook until golden brown, about 5 to 7 minutes total.

Serve hot with maple syrup and/or applesauce.

Cinnamon-Topped Raspberry Muffins

Makes 12 muffins

The sweet aroma of baking raspberry muffins will bring everyone into the kitchen.

1^1/$_2$ cups all-purpose flour
^{1}/$_2$ cup yellow or white cornmeal
^{1}/$_2$ cup plus 1 tablespoon sugar
2 teaspoons baking powder
^{1}/$_4$ teaspoon baking soda
^{1}/$_4$ teaspoon salt
^{1}/$_2$ teaspoon grated lemon peel
^{3}/$_4$ cup buttermilk
2 large eggs
^{1}/$_4$ cup melted butter or vegetable oil
1 cup raspberries
^{1}/$_2$ teaspoon ground cinnamon

Preheat the oven to 375 degrees.

Line 12 muffin cups (2^1/$_2$ inches) with paper liners or oil them. Combine the flour, cornmeal, ^{1}/$_2$ cup sugar, baking powder, baking soda, salt, and lemon peel in a large bowl. Set aside.

Combine the buttermilk, eggs, and butter in another bowl. Add to the flour mixture and stir just until combined. The batter will be thick. Fold in the raspberries. Divide the batter equally among the muffin cups. Combine the cinnamon with the remaining 1 tablespoon sugar and sprinkle over the top. Bake for 20 to 25 minutes, or until the muffins are golden brown and cooked through.

Serve warm.

Blackberry Sour Cream Coffee Cake with Streusel Filling

Makes 8 to 10 servings

Wonderful straight out of the oven, this coffee cake, layered twice with blackberries, can also be reheated.

STREUSEL

$1/2$ cup (packed) brown sugar

2 teaspoons all-purpose flour

2 teaspoons ground cinnamon

$1/2$ cup chopped, skinned hazelnuts or walnuts

SOUR CREAM BATTER

8 tablespoons (1 stick) butter, softened

$1^1/4$ cups granulated sugar

2 large eggs

1 cup sour cream

1 teaspoon vanilla extract

1 teaspoon grated lemon peel

$2^1/2$ cups all-purpose flour

1 teaspoon baking powder

$1/2$ teaspoon baking soda

$1/2$ teaspoon salt

$1^1/2$ to 2 cups fresh or frozen blackberries, rinsed if necessary (see Note)

To prepare the streusel, combine the brown sugar, flour, cinnamon, and nuts and mix well. Set aside.

To prepare the batter, cream the butter in the large bowl of an electric mixer until fluffy. Gradually beat in the sugar. Add the eggs, one at a time, beating well after each addition. Stir in the sour cream, vanilla, and lemon peel and mix well. Combine the flour, baking powder, baking soda, and salt and stir into the sour *cream mixture, mixing well.*

Preheat the oven to 350 degrees. Grease a 9-inch tube pan.

Combine the streusel with the blackberries. Layer one third of the batter into the pan. Sprinkle with half of the streusel, spacing the blackberries evenly. Repeat with another third of the batter, the remaining streusel, and the remaining batter.

Bake for about 1 hour, or until a cake tester or wooden pick inserted near the center comes out clean. Cool for 10 minutes on a rack. Turn out onto a platter.

Serve warm or at room temperature.

Note: If using frozen berries, thaw in an open dish for 15 to 20 minutes before using.

To Select Berries

When choosing, look for plump, fully colored berries that are firm-looking and fresh. A stained or wet container may be an indication that the fruit is overripe or that it has been sitting too long.

Once picked, none of these berries (strawberries, bramble berries, blueberries, gooseberries, currants, and cranberries) will continue to ripen. They will, however, deteriorate, especially if left at room temperature.

Strawberry Scones

Makes 8 scones

Sweet bits of strawberry or whole dried cranberries lend an appealing texture to these sprightly scones. To me, scones always seem to taste best when served warm from the oven, whether for breakfast or tea. Split the scones and add honey, if you like.

> 2 cups all-purpose flour
> ^{1}/4 cup (packed) brown sugar
> 2 teaspoons baking powder
> ^{1}/4 teaspoon salt
> 2 teaspoons grated fresh orange peel
> 4 tablespoons (^{1}/2 stick) chilled butter, cut into small pieces
> 1 cup diced fresh strawberries, whole raspberries, or
>   other sweet berries
> About ^{2}/3 cup milk

Preheat the oven to 425 degrees.

Combine the flour, brown sugar, baking powder, salt, and orange peel in a large bowl and mix well. Cut in the butter with a pastry blender or rub together with your fingers until crumbly. Stir in the strawberries. Gradually stir in enough milk to make a soft dough. Gather into a loose ball and knead 3 times on a lightly floured surface. Pat out to a 7-inch round, about ^{3}/4 inch thick. The dough will be sticky. Cut into 8 wedges and place ^{1}/2 inch apart on an ungreased baking sheet.

Bake for 20 to 25 minutes, or until golden brown.

Serve warm.

Variations: Substitute ^{1}/2 to ^{3}/4 cup dried cranberries for the strawberries. Substitute ^{3}/4 teaspoon grated fresh ginger for the orange peel.

Lingonberry Almond Braid

Makes 8 to 12 servings

A few years ago, at Camp Denali, in Alaska's Denali National Park, I discovered some tiny red, and very tart, lingonberries growing low to the ground. The following morning, camp baker Nancy Bale made a fragrant lingonberry yeast bread. She generously shared the recipe.

YEAST DOUGH

About 2^1/$_2$ cups all-purpose flour

3 tablespoons sugar

1 package active dry yeast

1 teaspoon salt

^{1}/$_2$ cup milk

3 tablespoons butter

1 large egg, slightly beaten

ALMOND FILLING

^{1}/$_4$ cup sliced almonds

2 ounces almond paste

1 tablespoon egg white, slightly beaten

^{1}/$_4$ cup apricot jam, warmed

^{1}/$_3$ cup lingonberries in sugar or cranberry sauce (see Note)

1 egg yolk beaten with 1 tablespoon water, for glaze

POWDERED SUGAR GLAZE

^{1}/$_3$ cup powdered sugar

1 teaspoon lemon juice

About ^{3}/$_4$ teaspoon hot water

^{1}/$_4$ cup toasted sliced almonds

To prepare the dough, combine 2 cups of the flour, the sugar, yeast, and salt in the large bowl of an electric mixer. Heat the milk and butter in a saucepan or microwave oven to between 120 and 130 degrees. Pour over the flour mix-ture and beat for 1 minute. Add the egg and beat for 2 minutes more at medium speed. Gradually stir in enough of the remaining flour to make a soft dough. Knead with the dough hook or by hand on a floured surface until smooth and elastic. Place in an oiled bowl, cover, and let rise in a warm place until doubled, 30 to 40 minutes.

To prepare the filling, whirl the sliced almonds in a food processor until finely chopped. Add the almond paste and process until well mixed. Stir in the egg white. Mix with the apricot jam. Set aside.

Punch down the dough and knead briefly on a floured surface. Cover and let rest for 10 minutes. Roll and stretch the dough to a 16 × 9-inch rectangle. Spread the filling lengthwise down the center in a 3-inch-wide strip. Spread the lingonberries over the filling. Using a pastry wheel or knife, make about 15 parallel cuts on an angle, about 1 inch apart down each side of the dough. Be-ginning at the top, alternately fold left and right strips at a slight angle over the filling, sealing the ends. Transfer to a greased baking sheet. Cover and let rise until almost double, about 45 minutes.

Preheat the oven to 375 degrees.

Brush the braid with the egg glaze. Bake for 30 to 35 minutes, or until golden brown. Cool on the pan 10 minutes on a rack.

Meanwhile, prepare the glaze. Stir together the powdered sugar, lemon juice, and enough hot water to make a spreading consistency. Spoon over the braid and sprinkle with the sliced almonds. Serve warm.

Alternatively, cool the unglazed braid completely. Reheat, covered loosely with foil, at 300 degrees for 25 minutes, or until thoroughly heated. Drizzle the glaze over the braid and sprinkle with the sliced almonds. Serve warm.

Note: Jars of wild Swedish lingonberries in sugar or syrup are available in many markets. If unavailable, whole-berry cranberry sauce makes a good alternative.

Cranberry-Orange Nut Bread

Makes 1 loaf

This moist bread keeps well. It's best to wrap it and let it stand overnight before cutting. The flavor of orange and hazelnuts goes well with the cranberries.

2 cups all-purpose flour
1^{1}/$_{2}$ teaspoons baking powder
^{1}/$_{2}$ teaspoon baking soda
^{1}/$_{2}$ teaspoon salt
1 egg, slightly beaten
2 tablespoons vegetable oil
^{3}/$_{4}$ cup orange juice
1 tablespoon and 2 teaspoons grated orange peel
1 teaspoon vanilla extract
1 cup sugar
1 cup fresh or frozen cranberries, rinsed and picked over, coarsely
   chopped (see Note)
^{1}/$_{2}$ cup coarsely chopped, skinned hazelnuts

Preheat the oven to 350 degrees. Grease a 9 × 5 × 3-inch loaf pan.

Combine the flour, baking powder, baking soda, and salt. Set aside. Beat together the egg, oil, orange juice, orange peel, vanilla, and sugar in a large bowl. Add the flour mixture and stir just until moistened. Stir in the cranberries and hazelnuts. Spoon into the pan.

Bake for 60 minutes, or until a wooden pick inserted near the center comes out clean. Cool in the pan on a rack for 15 minutes. Remove from the pan and cool completely. Wrap in plastic or foil and let stand for 12 hours or overnight before cutting.

Note: If the cranberries are frozen, chop them without thawing and add directly to the batter.

Olallie Berry Smoothie

Makes about 2 servings

This simple-to-put-together thirst quencher is great for breakfast or a mid-day cooler. Try other berries, such as raspberries, blueberries, or strawberries, adjusting the sugar or honey to taste.

> *1 cup fresh or frozen Olallie berries or other blackberries, rinsed if*
>   *necessary*
> *1 ripe banana, cut into pieces*
> *1 cup plain yogurt*
> *About 2 tablespoons sugar or honey*
> *1/2 cup crushed ice*

Blend the berries, banana, and yogurt in a blender or food processor, pulsing on and off until the mixture is smooth. Taste and add sugar or honey to taste. Add the crushed ice and blend until smooth.

Serve immediately.

To Refrigerate Berries

Sort the berries, discarding any mushy or moldy ones. If fragile, such as strawberries and raspberries, place the unwashed berries in a single layer on a paper towel-lined tray. Cover lightly with more paper towels or vented plastic wrap. Check daily: They should keep for a day or two.

Firmer berries, such as currants, gooseberries, and usually blueberries, will hold for a week or more, depending on when they were picked.

Cranberries can be refrigerated in their perforated plastic bags for up to two weeks.

Salads and Entrees

Summer Mixed Fruit Salad

Makes about 6 servings

Berries with melon are equally fitting at breakfast, as a salad, or—drizzled with a little brandy—for dessert. For an elegant presentation, line plates with butter lettuce and slices of Charentais melon and spoon the berry mixture on top.

2 cups strawberries, rinsed, hulled, and halved or quartered
1 cup mixed fresh berries, such as blackberries, blueberries, and/or raspberries
2 cups 1/2-inch cubes watermelon
1 cup cranberry juice cocktail
2 tablespoons sugar
1 to 2 tablespoons brandy or Triple Sec (optional)

Rinse the berries as necessary and pat to dry. Combine with the melon in a large bowl. Pour the cranberry juice over and sprinkle with sugar. Toss gently. Add the brandy, if using, and toss again. Refrigerate for up to 4 hours.

To serve, spoon the berry mixture into bowls.

Blueberry Salad with Blue Cheese

Makes 6 to 8 servings

This first-course salad goes well with a light white wine. Pair the blueberries with blue cheese from Oregon's Rogue River Valley Creamery or a creamy Italian gorgonzola and serve the salad with an Oregon pinot gris or an Italian pinot grigio.

BLUEBERRY VINAIGRETTE

$1/3$ cup extra virgin olive oil

2 tablespoons Blueberry-Basil Vinegar (page 92) or white wine vinegar

1 tablespoon lemon juice

1 tablespoon honey

1 teaspoon grainy mustard

$1/4$ teaspoon freshly ground black pepper

Dash of salt

$1^1/2$ cups blueberries, rinsed and picked over

3 cups (lightly packed) watercress sprigs

3 cups (lightly packed) spinach, torn into bite-size pieces

4 cups (lightly packed) gourmet salad greens or red leaf lettuce,
    torn into bite-size pieces

$1/3$ cup crumbled blue cheese

To prepare the blueberry vinaigrette, combine the ingredients in a small jar with a tightly fitting lid and shake well. Let stand at room temperature for up to 1 hour.

Place the blueberries in the bottom of a large salad bowl. Add the vinaigrette and toss. Layer the watercress, spinach, and gourmet salad greens on top. Refrigerate, covered with a damp paper towel, for up to 4 hours.

To serve, sprinkle the blue cheese over the greens and toss thoroughly.

Variation: Raspberries or small strawberries may be substituted for the blueberries. Use Raspberry Vinegar (page 93) or Strawberry Vinegar (page 94) as appropriate.

Raspberry Salad with Warm Pears

Makes 6 servings

Raspberries and pears share the same seasons. The first local raspberries ap-
pear in late June and continue well into the autumn. In the summer, use
Bartletts for this salad; in the fall, choose Anjou, Bosc, or Comice.

> 1 tablespoon and 2 teaspoons balsamic vinegar
> 1 tablespoon sugar
> $1/8$ teaspoon salt
> Freshly ground black pepper
> $1/4$ cup plus 1 tablespoon extra virgin olive oil
> 3 to 4 tablespoons minced shallots or onion
> $1/4$ teaspoon fresh or dried thyme
> 2 pears, cored and cut into bite-size pieces
> 10 to 12 cups torn mixed salad greens
> 1 to $1^1/2$ cups fresh or partially thawed frozen raspberries

Combine the vinegar with the sugar, salt, and pepper to taste. Whisk in
$1/4$ cup oil. Let stand at room temperature for up to 2 hours.

Heat the remaining 1 tablespoon oil in a skillet over medium heat. Add the
shallots and thyme and sauté until the shallots are tender. Add the pears and
cook, stirring gently, until warm, about 5 minutes.

Place the salad greens in a large bowl. Whisk the dressing, pour over the
greens, and toss. Divide the greens among the plates and arrange the pears and
raspberries on top.

Variation: Sliced strawberries may be substituted for the raspberries.

Strawberry and Shrimp Salad

Makes 4 main-dish servings

This spring or summer dish is quick and easy to put together. Sweet, aromatic strawberries and tangy shrimp enliven this colorful arranged salad.

STRAWBERRY MARINADE AND DRESSING

2 tablespoons Strawberry Vinegar (page 94) or
   white wine vinegar
1 tablespoon lemon juice
2 teaspoons honey
1 teaspoon Dijon mustard
1/4 teaspoon salt
1/4 teaspoon freshly ground black pepper
1/2 cup olive oil

1 pound large shrimp, peeled and deveined
2 heads butter lettuce, leaves separated, washed, and crisped
1 pound asparagus or green beans, trimmed
2 cups strawberries, rinsed, hulled, and halved if large

Prepare a charcoal fire or preheat the broiler.

To prepare the strawberry marinade and dressing, combine the vinegar, lemon juice, honey, mustard, salt, and pepper in a bowl. Gradually whisk in the olive oil.

Place the shrimp in a shallow dish and pour over 3 tablespoons of the dressing. Let marinate for 15 to 30 minutes, stirring occasionally.

Cook the asparagus in boiling water or in a microwave oven until crisp-tender, 2 to 3 minutes. Drain and cool.

Divide the large lettuce leaves among 4 plates. Shred the inner leaves and place on top.

Skewer the shrimp and discard any leftover marinade. Grill or broil, turning once, until opaque, about 6 to 8 minutes.

To serve, arrange the berries, asparagus, and shrimp on the lettuce and spoon the remaining dressing over.

Variation: Substitute raspberries and Raspberry Vinegar (page 93) for the strawberries and Strawberry Vinegar.

To Serve Berries

For the best flavor, remove the berries from the refrigerator an hour or two before serving. Pick over and remove any debris or soft, moldy, or immature berries. Then, at the last minute, rinse with a gentle spray of cool water and blot dry. For strawberries, rinse before you hull them (remove the caps), so the berries do not become waterlogged.

Raspberry Chicken Salad

Makes 2 main-dish servings

Raspberry vinegar, raspberry jam, and fresh raspberries lend real sparkle to this refreshing chicken salad. You can double or triple the recipe as you like.

RASPBERRY MARINADE AND DRESSING

3 tablespoons Raspberry Vinegar (page 93)
2 tablespoons raspberry jam
1 teaspoon olive oil
$^1/_4$ teaspoon Dijon mustard
$^1/_4$ teaspoon grated lemon peel
Salt and freshly ground black pepper

2 boneless and skinless chicken breast halves
3 to 4 cups torn mixed salad greens
$^1/_2$ cup raspberries
2 tablespoons goat cheese, crumbled

Prepare a charcoal fire or preheat the broiler.

To prepare the marinade and dressing, combine the vinegar, jam, oil, mustard, lemon peel, and a dash each of salt and pepper in a shallow dish. Stir until smooth.

Place the chicken breasts in a shallow dish, add a scant 2 tablespoons of the marinade, and turn to coat. Let stand for 5 to 15 minutes, turning once.

Coat the grill rack with vegetable oil cooking spray. Place the chicken 4 to 6 inches from the heat. Discard the marinade. Grill or broil, turning once, until the chicken is opaque when tested at its thickest part, about 10 to 12 minutes. Meanwhile, arrange the salad greens on 2 plates.

To serve, slice the cooked chicken at an angle and transfer each breast to a plate. Garnish with the raspberries and cheese. Spoon the dressing over the salads.

Grilled Salmon with Blueberry–Red Onion Relish

Makes 4 servings

The tart blueberry relish is a great foil for the richness of the salmon. Start the grill first, then prepare the relish and fish.

BLUEBERRY–RED ONION RELISH

3/4 cup finely diced red onion
1/2 teaspoon vegetable oil
1/2 cup orange juice
1 1/2 teaspoons red wine vinegar
3/4 cup blueberries, rinsed and picked over
1 teaspoon (packed) brown sugar
1/4 teaspoon grated orange peel
Dash of salt
1 teaspoon cornstarch

2 tablespoons orange juice
1/2 teaspoon vegetable oil
4 salmon fillets or steaks

Prepare a charcoal fire.

To prepare the relish, sauté the onion in the oil in a saucepan over medium-high heat until translucent. Add the 1/2 cup orange juice and vinegar and bring to a boil. Stir in the blueberries, brown sugar, orange peel, and salt. Reduce the heat and simmer for 10 to 15 minutes, or until the onion is just tender. Remove from the heat. Combine the cornstarch and 1 teaspoon water and stir into the hot berry mixture. Return the pan to the heat and bring to a boil. Remove from the heat. Cover and keep warm while cooking the salmon. (The relish can be covered and refrigerated for 24 hours, then reheated in a saucepan over low heat, stirring frequently, or in a microwave oven.)

Combine the 2 tablespoons orange juice and oil in a glass dish. Add the salmon and coat both sides. Marinate for 5 to 15 minutes.

Coat the grill rack with vegetable oil cooking spray. Place the fish on the grill rack 4 to 6 inches above medium-hot coals. DIscard the marinade. Grill, turning once, until the fish is opaque and flakes when tested at its thickest point, about 10 minutes for each inch of thickness.

To serve, transfer to plates and spoon the Blueberry–Red Onion Relish over the salmon.

Pork Roast with Blackberry–Pinot Noir Sauce

Makes about 6 servings

A lively blackberry–pinot noir sauce glistens over the sliced pork. Serve this appealing entree with an Oregon pinot noir or a zinfandel from California's Dry Creek Valley.

1 boneless pork loin roast (3 to 3^1/$_2$ pounds)
1 teaspoon olive oil (optional)
^{1}/$_4$ teaspoon dried thyme

BLACKBERRY–PINOT NOIR SAUCE

1 cup fresh or partially thawed frozen unsweetened blackberries, rinsed if necessary
^{1}/$_2$ cup low-salt chicken broth, canned or homemade
4 tablespoons (^{1}/$_2$ stick) plus 1 tablespoon butter
1 large shallot, minced
^{1}/$_2$ cup pinot noir, zinfandel, or other full-bodied red wine
1 tablespoon red wine vinegar
1 teaspoon sugar
^{1}/$_4$ teaspoon salt
^{1}/$_4$ teaspoon freshly ground black pepper

Sprigs of watercress
^{1}/$_2$ cup fresh blackberries, for garnish

Preheat the oven to 350 degrees.

Place the roast, fat side up, on a rack in a shallow baking dish. If the roast is lean, brush with olive oil. Sprinkle with thyme. Roast for 1 hour and 45 minutes, or until an instant-read thermometer inserted in the center of the roast reads 160 degrees for well done. Transfer the roast to a platter. Cover loosely with foil and let stand for 15 minutes before slicing.

Meanwhile, make the blackberry–pinot noir sauce. Combine the berries and broth in a food processor. Pulse on and off until liquefied. Strain out the seeds and reserve the berry pulp. Melt the 1 tablespoon butter in a saucepan over medium-low heat. Add the shallot and sauté for 5 minutes, or until soft. Stir in the blackberry-broth mixture, the pinot noir, vinegar, sugar, salt, and pepper. Cook over high heat, stirring occasionally, until the mixture is reduced by half, $^3/_4$ to 1 cup, about 5 minutes. (The sauce can be held for up to 1 hour. Reheat if necessary.) Whisk in the remaining 4 tablespoons butter, 1 tablespoon at a time. Taste and season, if necessary, with salt and pepper. Keep warm.

To serve, slice the pork roast and arrange on warm dinner plates. Spoon some of the sauce over each serving or pass the sauce in a gravy boat. Garnish with the watercress and fresh blackberries.

Red Currant Sauce for Warm or Cold Meats

Makes about ^{2}/3 cup

English-born Barbara Haney makes this fresh currant sauce with red currants from her garden. The sauce may be served warm or cold, with venison, lamb, beef, turkey, or pheasant. Barbara suggests adding a couple of table-spoons of meat juices to the sauce before serving.

1 orange
1 small lemon
1 cup fresh or frozen red currants
^{1}/4 cup and 1 teaspoon sugar
2 tablespoons port
^{1}/2 teaspoon ground ginger
^{1}/2 teaspoon Dijon mustard
Salt and freshly ground black pepper

Wash and coarsely grate the colored part of the peel from the orange and lemon. Place the peel in a saucepan with cold water to cover. Bring to a boil, reduce the heat, and simmer for about 4 minutes. Drain well and set aside.

Squeeze the juice from the orange and lemon. You should have about ^{1}/2 cup juice. Combine the citrus juice and currants in a saucepan. Bring to a boil, reduce the heat, and simmer, uncovered, until soft, 10 to 12 minutes. Stir in the reserved citrus peel, sugar, port, ginger, mustard, and salt and pepper to taste. Cook at a low boil, stirring frequently, for 10 to 20 minutes, or until reduced to a little more than ^{3}/4 cup. Serve warm or pour into a container with a tightly fitting lid and refrigerate for up to several weeks. Reheat before serving, if desired.

Variation: Fresh or frozen cranberries can be substituted for the currants.

Cumberland Sauce for Game, Beef, or Poultry

Makes about ³/4 cup

Another easy recipe from Barbara Haney. This one is adapted from one by the English cookery writer, the late Jane Grigson. The sauce can be served warm with venison, lamb, or beef or at room temperature with turkey or pheasant. This recipe can be doubled.

> 1 orange
> 1 small lemon
> ¹/2 cup red currant jelly
> 2 tablespoons port
> ¹/2 teaspoon Dijon mustard
> ¹/2 to ³/4 teaspoon ground ginger or grated fresh ginger
> Salt and freshly ground black pepper

Wash and coarsely grate the colored part of the peel from the orange and lemon. Place the peel in a saucepan with cold water to cover. Bring to a boil, reduce the heat, and simmer for about 4 minutes. Drain well and set aside.

Squeeze the juice from the orange and lemon. You should have about ¹/2 cup juice. Heat the currant jelly over low heat until it melts. Stir in the citrus juice, reserved citrus peel, port, mustard, ginger, and salt and pepper to taste. Simmer for 15 minutes.

Serve warm or pour into a container with a tightly fitting lid and refrigerate for up to several weeks. Bring to room temperature before serving.

BERRIES IN THE KITCHEN

BERRY	TO KEEP	NUTRITION	COMMENTS
Blackberry	Perishable; refrigerate.	1 cup = 74 cal., 7 gr. fiber.	If using frozen, add 1 Tbsp. flour to the batter, cook 5 min. longer.
Blueberry and Huckleberry	Refrigerate for up to 2 weeks.	1 cup = 82 cal.; high in vit. A, vit. C, calcium, phosphorus, and potassium.	If using frozen, cook 10 min. longer.
Cranberry	Refrigerate in perforated plastic bags up to 2 weeks.	1 cup = 46 cal.; high in vit. and vit. C; rich in pectin.	Berries start to decompose if wet.
Currant	Refrigerate for up to one week.	1 cup = 70 cal.; high in vit. C and fiber.	Use a fork to comb off the stems.
Gooseberry	Refrigerate for up to one week.	1 cup = 65 cal.; high in vit. C and fiber.	Remove blossom and stem ends.
Jostaberry	Refrigerate for up to one week.	1 cup = 65 cal.; high in Vit. C and fiber.	A dark skin indicates ripeness.
Logan, Boysen, Tayberry, etc.	Refrigerate for up to one week.	1 cup = 74 cal., 7 gr. fiber.	If using frozen, add 1 Tbsp. flour to the batter, cook 5 min. longer.
Raspberry	Very perishable; refrigerate.	1 cup = 61 cal., 6 gr. fiber.	If using frozen, add 1 Tbsp. flour to the batter, cook 5 min. longer.
Strawberry	Very perishable; refrigerate for 1–2 days.	1 cup = 45 cal.; high in vit. C, iron, and fiber.	Rinse, then hull (remove caps).

Condiments and Preserves

Cranberry and Pear Relish

Makes about 4 cups

Tangy cranberries and juicy pears make a complementary duo for this chunky relish. Serve with poultry, ham, cold meats, or curry. This recipe can be doubled.

1 bag (12 ounces) fresh or frozen cranberries,
   rinsed and picked over
1 cup sugar
$3/4$ cup water
2 tablespoons golden raisins
1 tablespoon minced onion
$1/4$ teaspoon minced fresh ginger
1 cinnamon stick (about 2 inches)
$1/8$ teaspoon ground cloves
$1/8$ teaspoon salt
2 medium firm-ripe pears, such as Anjou or Bosc, rinsed

Combine all the ingredients, except the pears, in a $2^1/2$- to 3-quart nonreactive saucepan. Bring to a boil, reduce the heat, and simmer for 15 minutes, stirring occasionally.

Meanwhile, core the pears and cut into $1/2$-inch pieces. You should have about $2^1/2$ cups. Add the pears to the saucepan and continue to simmer, stirring occasionally, until heated through, about 8 minutes more. Remove the cinnamon stick.

Pour into a serving bowl and let cool. Or pour the hot relish into hot clean jars. Cover and refrigerate for up to 1 month. Freeze to keep longer.

Blueberry-Basil Vinegar

Makes about 2 cups

Capture the essence of summer with the pungent fragrance of basil and the sweet taste of blueberries. Make a double batch for holiday gifts.

> 1^1/4 cups crushed blueberries
> ^{1}/4 cup (firmly packed) torn basil leaves
> 2 cups distilled white vinegar
> 1 teaspoon sugar

Combine the crushed berries and basil in a glass measure or bowl. Stir to mix and set aside.

Combine the vinegar and sugar in a nonreactive saucepan and bring to a boil. Pour the hot vinegar over the berry mixture. Cover the measure or bowl and let stand in a cool place for 4 days or up to 2 weeks.

Scald two 1-cup vinegar bottles. Strain the berry-vinegar mixture through a fine sieve or several layers of cheesecloth into a clean glass measure. Pour the vinegar into the clean bottles and tightly cork or cap.

Raspberry Vinegar

Makes about 4 cups

For the past several years, I've been the lucky recipient of a bottle of ruby-red raspberry vinegar on my birthday. My friend Betty Orians makes this special vinegar every year, usually in July, using berries from her garden. She notes that although the vinegar will keep for years, the younger it is, the fruitier the flavor.

The quantity of sugar can vary from none to four cups. A sweet vinegar can be sparingly added to seltzer water, club soda, or white wine. The unsweetened or lightly sweetened vinegar makes an excellent oil-and-vinegar salad dressing.

2 quarts red raspberries
1 quart red wine vinegar, cider vinegar, or distilled white vinegar
2 quarts red raspberries, to be used 2 days later
0 to 4 cups sugar

Place 2 quarts of the raspberries in a large glass or ceramic bowl. Pour the vinegar over, cover with plastic wrap, and set in a cool place, such as a basement, for 2 days.

Strain the mixture in a colander set over a bowl, pressing lightly. Discard the seeds and pulp and pour the liquid into a clean bowl. Add the remaining 2 quarts berries. Cover and set in a cool place for 2 days longer.

Strain the berry-vinegar mixture through a triple thickness of cheesecloth into a 3-quart nonreactive pan. For a clear liquid, avoid squeezing. Add the sugar, if using, and bring to a boil, stirring until the sugar is dissolved. Reduce the heat and simmer for 10 minutes. Remove from the heat and skim off any foam. Let cool.

Meanwhile, scald several 1-cup or 2-cup vinegar bottles. Pour the liquid into the clean bottles and tightly cork or cap.

Strawberry Vinegar

Makes about 2 cups

Splash this vinegar over whole berries for a refreshing snack. Or stir a spoonful into sparkling water with crushed ice and garnish with a whole strawberry for a colorful spritzer.

> 5 cups (about 1¹/₂ pounds) strawberries,
>   rinsed, hulled, and sliced
> 1 cup white wine vinegar
> 1¹/₂ teaspoons sugar

Combine the berries and vinegar in a glass measure or bowl. Cover with plastic wrap and let stand at room temperature for 24 hours, stirring occasionally.

Scald two 1-cup vinegar bottles. Strain the berry-vinegar mixture through a fine sieve or several layers of cheesecloth into a large nonreactive saucepan, pressing lightly. Discard the pulp and seeds. Stir in the sugar and bring to a boil. Simmer, covered, for 2 minutes. Remove from the heat and let cool. Pour the vinegar into the clean bottles and tightly cork or cap.

Jeweled Cranberry Confit

Makes about 2 cups

A gem of a recipe. Anne Willan, the founder of La Varenne Cooking School, gave me this years ago. It is nice and tart, not like some of those overly sweet cranberry sauces.

> 1 tablespoon plus 2 teaspoons butter
> 1 bag (12 ounces) cranberries, rinsed and picked over
> ³/₄ cup sugar

Preheat the oven to 350 degrees.

Spread the 2 teaspoons of butter in a large shallow baking pan, such as a 13 × 9 × 2-inch glass dish. Add the cranberries and sprinkle with the sugar. Dot with the remaining 1 tablespoon of butter. Cover with foil and bake for 35 to 45 minutes, or until berries are soft. Uncover and bake for 15 minutes more, or until the berries are transparent and any liquid has evaporated. Serve warm or reheat.

To Freeze Berries

Remove the stems and blossom ends from the fruit as needed. Rinse berries quickly with cool water, blot dry, and place in a single layer, without touching, on a rimmed baking pan. Freeze until firm, then transfer to freezer bags or covered storage containers.

Strawberries tend to become mushy after freezing. An alternate method is to halve or slice the berries, sprinkle with sugar, and freeze, as above, to use for desserts.

Since cranberries are available only in the fall, consider freezing a bag or two. Place them in a freezer bag and freeze for up to one year.

Cranberry, Orange, and Lemon Marmalade

Makes 3¹/2 to 4 cups

Like a burst of sunshine on a cold winter's morning, this rosy marmalade will surely wake you up. Serve it on breakfast toast or with yeast rolls for Thanksgiving dinner.

2 large oranges
2 large lemons
1 cup fresh or frozen cranberries, rinsed and picked over
1¹/2 cups sugar

Wash and cut the oranges and lemons lengthwise in half through the stem ends. Place the fruit flat side down and cut lengthwise in half again. Thinly slice the fruit crosswise into slices ¹/8 inch or less. Discard the ends and seeds. You should have 5 to 6 cups.

Combine the orange and lemon slices and 1 cup of water in a large heavy saucepan and bring to a boil. Reduce the heat and cook at a low boil, stirring occasionally, for 20 minutes or until almost tender. Add the cranberries and boil gently for 10 minutes, stirring to avoid scorching. Stir in the sugar and return to a boil. Cook, stirring frequently, until slightly thickened, for 10 to 15 minutes. The mixture will continue to thicken as it cools.

Ladle into clean half-pint jars, cover, and cool. Store in the refrigerator for up to 4 weeks.

Variation: In summer, substitute fresh gooseberries for the cranberries.

Blackberry Jam-Jell

Makes about 8 cups

My mother-in-law showed me how to capture the aroma of a warm summer day of berry picking in a jar of seedless jam. And all without the hassle of a jelly bag.

Have the jars ready before you start. I don't bother with sterilizing. I just keep the jam-jell in the refrigerator. It goes fast enough as is.

> *3 to 4 quarts blackberries, rinsed if necessary*
> *About 5 cups sugar*
> *About $^1/_3$ cup lemon juice*

Place the berries in an 8-quart kettle. Crush with a potato masher. Heat to a boil. Work the berry pulp through a sieve or food mill into a 2-quart measure or large bowl and discard the seeds. Clean the kettle.

Measure the volume of puree. (You should have about 5 cups.) Add an equal volume of sugar. Stir in the lemon juice to taste. Return the mixture to the clean kettle and heat to 220 degrees on a candy thermometer, or until the jam-jell sheets off a spoon.

Ladle into hot half-pint jars. Seal, cool, and refrigerate until used.

Alternatively, sterilize the jars, fill and seal them, and place in a hot-water bath. Bring to a boil, and process for 10 minutes. Remove the jars from the water and cool on a rack.

Microwave Strawberry Jam

Makes about 1 cup

Carolyn Bertsch has made this simple jam for at least twenty years, starting back when she was first teaching microwave classes. She says that the little bit of butter adds to the flavor.

> 2 cups ripe strawberries
> $^1/_2$ cup sugar
> 1 to 2 tablespoons lemon juice
> $^1/_4$ teaspoon butter

Rinse, hull, and crush the strawberries. You should have about 1 cup. Stir together the strawberries, sugar, lemon juice to taste, and butter in a 2-quart glass measure or casserole. Microwave, uncovered, at high (100 percent) for 4 minutes. Stir well. Microwave for 4 minutes more and stir again. Test the jam by dropping a little off the side of a spoon onto a saucer. The droplets should be thick. Or wait 2 minutes and run your finger through the jam on the saucer; it should leave a clean streak. If the jam is too thin, microwave for 1 or 2 minutes more.

BERRIES, BERRIES, BERRIES,			
BERRY	**NW SEASON**	**TO SELECT**	**HOW SOLD**
Blackberry	Aug.–Late Sept.	Firm berries, free of moisture.	Half-pint, pint
Blueberry and Huckleberry	Late June-Late Sept.	Firm, plump berries with deep blue color.	Half-pint, pint
Cranberry	Mid-Sept.–Late Dec.	Firm, plump berries; avoid soft or shriveled fruit.	12-ounce bags
Currant	June–Mid-Aug.	Firm, dry berries.	Half-pint, pint
Gooseberry	May–Late Aug.	Firm, dry berries.	Half-pint, pint
Jostaberry	June–Mid-Aug.	Firm, dry berries.	Half-pint, pint
Logan, Boysen, Tayberry, etc.	Mid-June–Late Aug.	Firm, plump berries, free of moisture.	Half-pint, pint
Raspberry	June–Late Sept.	Firm, not soggy, free of moisture.	Half-pint, pint
Strawberry	June–Aug.	Plump, well rounded berries, fresh green caps.	Pint, quart, by the pound

Cakes

Summer Blueberry and Peach Upside-Down Cake

Makes 8 servings

I like to make a big circle of peach slices in the skillet and fill in with lots of blue-berries. The finished cake looks like a giant summer sunflower. If you are lucky enough to find huckleberries, this cake is even more yummy—if possible—with them.

8 tablespoons (1 stick) butter, softened
$^1/_2$ cup (packed) brown sugar
2 large peaches, peeled, pitted, and cut into 8 slices each
1 cup fresh or partially thawed frozen blueberries,
   rinsed and picked over if necessary
$^1/_2$ cup granulated sugar
1 large egg
1 teaspoon vanilla extract
1$^1/_4$ cups all-purpose flour
1$^1/_4$ teaspoons baking powder
$^1/_4$ teaspoon salt
$^1/_2$ cup milk

Melt 4 tablespoons of the butter in an 8- to 9-inch cast-iron skillet. Add the brown sugar and stir for 3 minutes over medium-low heat until bubbly. Carefully add the peach slices to the liquid in the pan. Cook for about 5 minutes, turning occasionally. Arrange the peach slices in a circle or other attractive design. Arrange the blueberries in the center and around the edge. Set aside.

Preheat the oven to 375 degrees.

Cream the remaining 4 tablespoons butter in the large bowl of an electric mixer. Gradually beat in the granulated sugar. Add the egg and vanilla. Combine the flour, baking powder, and salt and add to the egg mixture alternately with the milk. Spoon the batter evenly over the fruit.

Bake for 25 to 30 minutes, or until the top is golden and the cake springs back when pressed in the center. Cool in the pan for 5 minutes. Loosen the edges and invert onto a large platter. Serve warm or at room temperature.

Winter Cranberry and Pear Upside-Down Cake

Makes 8 servings

Ruby-red cranberries sparkle among the pear slices on top of the cake. Use firm-ripe Anjou or Bosc pears. They are ready when the stem end yields when gently pressed.

4 tablespoons ($^1/_2$ stick) plus 2 tablespoons butter, softened
$^1/_3$ cup (packed) brown sugar
2 small to medium firm-ripe pears, peeled,
   cored, and cut into thin slices
$^2/_3$ cup fresh or partially thawed frozen cranberries,
   rinsed and picked over
$1^1/_4$ cups all-purpose flour
$1^1/_2$ teaspoons baking powder
$^1/_2$ teaspoon salt
$^1/_2$ teaspoon ground cinnamon
$^1/_2$ teaspoon ground ginger
$^1/_4$ teaspoon ground nutmeg
$^2/_3$ cup granulated sugar
1 large egg, beaten
$^1/_2$ cup milk
$1^1/_2$ teaspoons grated lemon peel
Vanilla ice milk or frozen yogurt (optional)

Preheat the oven to 350 degrees.

Place the 2 tablespoons of butter in a 9-inch-round cake pan. Heat in the oven for 2 or 3 minutes, or until melted. Sprinkle with the brown sugar and arrange the pear slices in the pan, slightly overlapping. Fill in with cranberries and set aside.

Stir together the flour, baking powder, salt, cinnamon, ginger, and nutmeg and set aside. In the bowl of an electric mixer, beat the granulated sugar and the remaining 4 tablespoons of butter until well blended. Stir in the egg, milk, and lemon peel. Stir in the flour mixture. Pour the batter evenly over the fruit.

Bake for about 35 minutes, or until a wooden pick inserted near the center comes out clean. Cool for 5 minutes on a rack. Loosen the cake from the sides of the pan and invert onto a platter. Serve warm or at room temperature with vanilla ice milk or frozen yogurt, if desired.

Italian Almond Cake with Balsamic Strawberries

Makes 8 servings

Embellish this simple almond cake with the inspired combination of fresh strawberries, balsamic vinegar, and a few grinds of black pepper.

> $3/4$ cup (about $3^1/2$ ounces) finely ground almonds with skins
> $2/3$ cup sugar
> $1/4$ cup and 2 tablespoons all-purpose flour
> 1 teaspoon grated lemon peel
> $1/8$ teaspoon salt
> 4 large egg whites

STRAWBERRY TOPPING

> $4^3/4$ cups (about $1^1/2$ pounds) strawberries, rinsed, hulled, and
>   sliced
> $1/4$ teaspoon freshly ground black pepper
> 1 to 2 tablespoons balsamic vinegar
> 2 to 3 tablespoons sugar

Preheat the oven to 350 degrees. Coat a 9-inch springform pan with vegetable oil cooking spray.

Combine the almonds, sugar, flour, lemon peel, and salt. Beat the egg whites in a large bowl until stiff. Fold in the almond mixture, adding a little at a time. Spoon the batter into the pan and smooth to an even thickness.

Bake for about 25 minutes, or until golden brown on top and a wooden pick inserted near the center comes out clean. Cool for 10 minutes on a rack. Remove the sides of the pan and let the cake cool completely.

Toss the strawberries with the pepper, vinegar, and sugar to taste. Let stand at room temperature for 20 to 30 minutes.

To serve, cut the cake into wedges and place on dessert plates. Spoon the berries and collected juices over the top.

Blueberry Lemon Tea Cake

Makes 1 loaf

The affinity of blueberries and lemon is well known. This simple tea cake can serve as a breakfast bread, a hiker's take-along treat, or, with a dollop of whipped cream and a few more berries, a sophisticated dessert.

Fresh lemons are a must for this recipe. Grate just the bright yellow skin, not into the white pith.

> 1^1/2 cups all-purpose flour (see Note)
> 1 teaspoon baking powder
> ^{1}/2 teaspoon salt
> 2 tablespoons grated lemon peel
> 1 cup sugar
> 4 tablespoons (^{1}/2 stick) butter, softened
> 2 tablespoons lemon juice
> 2 large eggs
> ^{1}/2 cup milk
> 1^1/4 cups fresh or partially thawed frozen blueberries, rinsed,
>   blotted dry, and picked over if necessary

LEMON GLAZE

> ^{1}/4 cup lemon juice
> ^{1}/4 cup sugar

Preheat the oven to 350 degrees. Grease a 9 × 5 × 3-inch loaf pan.

Combine the flour, baking powder, and salt. Stir in the lemon peel and set aside. In a large bowl cream together the sugar, butter, and the 2 tablespoons lemon juice until fluffy. Add the eggs and mix until smooth. Stir in the dry ingredients alternately with the milk and mix just until moistened. Stir in the berries. Spoon into the pan.

Bake for 50 to 60 minutes, or until the cake is golden and a wooden pick inserted near the center comes out clean. Place the loaf, still in the pan, on a cooling rack and poke holes with a small skewer.

To prepare the glaze, combine the lemon juice and sugar. Drizzle over the loaf. Let cool for 30 minutes.

Remove the cake from the pan and place right side up on a platter. Serve warm or cool.

Note: Add 1 tablespoon more flour if using frozen berries.

Variations: Substitute $^1/_2$ to $^3/_4$ cup dried blueberries for the fresh blueberries. Blackberries may be used instead of blueberries.

To Thaw Berries

Place frozen berries in a colander and spray with cool water, just until the ice melts. The berries will still be hard in the center. Pat dry with paper towels.

Thawed berries will have more moisture than fresh. To keep the fruit from sinking when baked, toss the partially thawed berries with an extra tablespoon of flour before adding to muffin or cake batters. When baking a pie or cobbler, increase the cooking time by ten minutes or so, to allow the excess liquid to evaporate.

Strawberry Shortcake

Makes 6 servings

Shortcakes are easy, since the biscuits can be made well ahead. And remember, strawberries, like other summer berries, are most flavorful when at room temperature.

6 cups strawberries, rinsed, hulled, and sliced, or other berries
2 tablespoons crème de fraise strawberry liqueur (optional)
5 tablespoons sugar

BISCUITS

1³/4 cups sifted all-purpose flour
2³/4 teaspoons baking powder
¹/2 teaspoon salt
5 tablespoons chilled butter
1 cup milk

1 cup chilled whipping cream
1 tablespoon sugar
¹/4 teaspoon vanilla extract

Combine the strawberries, liqueur if using, and sugar to taste. Let stand while making the biscuits.

Preheat the oven to 450 degrees. Lightly grease a baking sheet.

Combine the flour, baking powder, and salt in a large bowl. Cut in the butter with a pastry blender or rub together with your fingers until the mixture is the consistency of cornmeal. Add the milk and stir just until the dough holds together. Dollop the dough onto the baking sheet in 6 equal portions. Bake for 17 to 20 minutes, or until golden. Serve warm or cool on a rack.

Whip the cream to soft peaks and add the sugar and vanilla. Split the biscuits and spoon the berries over the lower halves. Divide the cream over the berries and replace the biscuit tops. Serve at once.

Raspberry Brownies

Makes 16 squares

These brownies, topped with raspberries and raspberry jam, taste rich and fudgy, but they are surprisingly low in calories.

1^1/4 cups sugar
^{1}/4 cup vegetable oil
4 large egg whites
2 teaspoons vanilla extract
1 cup all-purpose flour
^{1}/2 teaspoon baking powder
^{1}/4 teaspoon salt
^{1}/4 cup unsweetened cocoa powder
^{1}/2 cup seedless raspberry jam, melted in a
    microwave oven or small saucepan
32 (about 1^1/4 cups) fresh raspberries, halved

Preheat the oven to 350 degrees. Coat a 9-inch-square pan with vegetable oil cooking spray.

Combine the sugar and oil in a bowl. Beat with an electric mixer on medium speed. Add the egg whites and vanilla and beat well. Stir together the flour, baking powder, salt, and cocoa. Add to the sugar mixture and beat well. Spread the batter in the pan.

Bake for 23 minutes, or until a wooden pick inserted near the center comes out clean. Cool for 20 minutes on a rack. Spread with half of the warm jam. Place the raspberries, cut side down, on top. Spread with the remaining jam. Cool completely. Cut into 16 2-inch squares. Serve at room temperature. Refrigerate any remaining brownies after 3 hours.

Red, White, and Blueberry Cheesecake Bars

Makes 24 pieces

Jane Morimoto, Vice President and Director of Test Kitchens for the Evans Food Group in Seattle, says you can replace the fresh berries and cranberry juice with a can of blueberries, if you like.

GRAHAM CRACKER CRUST

1 cup graham cracker crumbs
$1/4$ cup sugar
4 tablespoons ($1/2$ stick) melted butter

CHEESECAKE FILLING

2 large eggs
8 ounces low-fat cream cheese, completely softened
$1/2$ cup sugar
$1/2$ teaspoon vanilla extract

BLUEBERRY TOPPING

$1/2$ cup cranberry juice
$1/4$ cup sugar
2 tablespoons cornstarch
1 tablespoon lemon juice
$2^1/2$ cups fresh blueberries, rinsed and picked over,
    or frozen berries (see Note)
Whipped cream (optional)

Preheat the oven to 375 degrees.

To prepare the crust, combine the graham cracker crumbs, sugar, and butter in a 13 × 9 × 2-inch pan and mix well. Press into an even layer in the bottom of the pan. Bake for 2 minutes. Cool on a rack.

To make the filling, beat the eggs, cream cheese, sugar, and vanilla with an electric mixer until smooth. Spoon into the pan. Bake for about 15 minutes, or until firm in the middle when gently shaken. Cool on a rack.

To prepare the topping, combine the cranberry juice and sugar in a sauce-pan. Heat over medium heat until simmering. Dissolve the cornstarch in $^1/_4$ cup water and stir into the cranberry juice mixture. Cook and stir until thickened and clear, then simmer for 1 minute. Add the lemon juice and let cool. Stir in the blueberries. Spread the topping over the filling. Refrigerate for 1 hour, or until thoroughly chilled.

To serve, cut into 2-inch squares. Top each piece with a dollop of whipped cream, if desired.

Note: To use frozen blueberries, rinse the frost off the berries under cold run-ning water and blot dry.

Variation: Substitute one 15- to 16-ounce can of blueberries for the fresh or frozen berries, cranberry juice, and sugar. To make the topping, pour the syrup from the canned blueberries into a saucepan. Add the 2 tablespoons cornstarch and whisk to dissolve. Cook and stir until thickened and clear, then simmer for 1 minute. Add the lemon juice and let cool. Stir in the blueberries and continue as directed above.

Berry Trifle

Makes 12 or more servings

An attractive, easy-to-make dessert that serves a crowd. Summer blueberries, strawberries, and raspberries make a complementary trio in this English-inspired dish. Substitute a purchased sponge cake, angel food cake, or pound cake, if you like.

> *1 recipe Sponge Cake (page 112)*
> *3 cartons (8 ounces each) low-fat vanilla yogurt*
> *1 package (10 ounces) frozen raspberries in light syrup, thawed*
> *3 cups (about 1 pound) fresh strawberries, rinsed, hulled, and halved or quartered*
> *2 cups fresh blueberries or 1 package (12 ounces) frozen blueberries, partially thawed*
> *3 tablespoons cream sherry or port*
> *Sliced strawberries and/or whole blueberries, for garnish*

Prepare the Sponge Cake (see page 112).

Place a strainer lined with 4 layers of cheesecloth over a bowl. Spoon the yogurt into the strainer and let drain for 1 hour. Discard the liquid and spoon the yogurt into a bowl. Cover and refrigerate.

Puree the raspberries in a blender or food processor. Pulse on and off until smooth. Transfer to a large bowl and stir in the strawberries and blueberries. Set aside.

Line the bottom of a 2- or 3-quart glass bowl with one third of the cake cubes. Drizzle with 1 tablespoon of the sherry. Spoon half of the berry mixture over the cake cubes, arranging the fruit out to the edge of the bowl. Spoon one third of the yogurt mixture over the berry layer. Repeat the layers of cake, sherry, berries, and yogurt. Place the remaining cake cubes over the yogurt and drizzle with the remaining sherry. Spoon the remaining yogurt on top in an even layer.

Cover and chill for 4 hours or up to 24 hours.

Arrange sliced strawberries and/or whole blueberries on the trifle before serving.

Sponge Cake

Makes one 15 × 10 × 1-inch cake

1 cup sifted cake flour
1 teaspoon baking powder
^{1}/4 teaspoon salt
3 eggs, separated
^{3}/4 cup plus ^{1}/4 cup sugar
^{1}/4 cup lukewarm water
2 teaspoons vanilla extract

Preheat the oven to 350 degrees. Coat a 15 × 10 × 1-inch jelly-roll pan with cooking spray. Line with wax paper and spray again. Set aside.

Sift together the flour, baking powder, and salt. Set aside. In the large bowl of an electric mixer, beat the egg whites until soft peaks form. Gradually add ^{1}/4 cup sugar and beat until stiff but not dry. Transfer to a bowl and set aside. In the same mixer bowl, beat the egg yolks on medium speed for 1 minute. Gradually beat in the remaining ^{3}/4 cup sugar until the mixture is light and pale yellow, about 5 minutes. Add the water and vanilla, beating at low speed until blended. Gradually add the flour mixture to the egg yolk mixture, beating on low speed. Remove from the mixer and stir one quarter of the egg whites into the batter. Fold in the remaining egg whites. Pour the batter evenly into the pan.

Bake for about 15 minutes, or until golden and the top springs back when touched. (The cake may be tacky.) Loosen the edges and invert onto a rack. Peel off the wax paper and cool completely. Cut the cake with a serrated knife into ^{3}/4-inch cubes.

Pies and Tarts

Wild Blackberry Lattice Pie

Makes one 9-inch pie

A couple of summers ago, my sister Patricia and her family arrived with a cooler full of hand-picked Oregon wild blackberries. We decided on a pie with a lattice topping, so my niece Steffie could see how to do it. The first thing she bought after her visit was a zigzag pastry wheel.

TWO-CRUST PIE DOUGH

2 cups all-purpose flour

$^1/_2$ teaspoon salt

10 tablespoons (1$^1/_4$ sticks) chilled butter, cut into small pieces, or $^2/_3$ cup shortening

6 to 7 tablespoons ice water

BLACKBERRY FILLING

1 cup sugar

$^1/_3$ cup all-purpose flour

5 cups blackberries, rinsed if necessary

2 tablespoons milk

$^1/_4$ teaspoon ground cinnamon mixed with 2 teaspoons sugar

Vanilla ice cream (optional)

To prepare the dough, combine the flour and salt in a bowl. Cut in the butter with a pastry blender or rub together with your fingers until the mixture is crumbly, the size of small peas. Gradually sprinkle in the water, 1 tablespoon at a time, until the dough holds together when gathered with a fork. Press together into 2 disks, wrap in plastic, and refrigerate for at least 20 minutes or up to 24 hours. If made ahead, let stand at room temperature for 30 minutes before rolling out.

Roll out one pastry disk on a lightly floured surface to a 12-inch circle. Fold into quarters and transfer to a 9-inch pie plate. Unfold and ease the pastry into the pan. Trim the edge, allowing a 1-inch overhang.

To prepare the filling, combine the sugar and flour in a large bowl. Add the berries and toss. Spoon into the pastry-lined pie plate.

Preheat the oven to 400 degrees.

Roll out the second pastry disk to a 10-inch circle. Cut with a zigzag pastry cutter or knife into $^1/_2$-inch- or $^3/_4$-inch-wide strips. Place 1 long strip vertically across the center of the berry filling, and then add another strip horizontally. Place the next 2 longest strips parallel to the first strip. Add 2 more strips, parallel to the second, going under the first strip. Continue adding strips, two at a time, working from the center out and weaving the strips into a lattice. Trim the ends of the lattice strips to $^1/_2$ inch. Fold the lower pastry up over the top pastry and flute to seal. Brush the lattice with milk and sprinkle with the cinnamon sugar. Place the pie plate on a baking sheet to catch any drips.

Bake for 40 to 45 minutes, or until the fruit is bubbling (about 160 degrees on an instant-read thermometer).

Serve the pie warm or at room temperature, with a scoop of vanilla ice cream, if desired.

Eileen's Blueberry-Crowberry Pie

Makes one 9-inch pie

Eileen, my sister who lives in Alaska, is an expert berry picker. During one of my visits, she transformed tiny crowberries and not much bigger blueberries into a tart-sweet pie. If you don't have a source of crowberries, which is likely since they grow primarily in Alaska, blueberries or huckleberries will do just fine.

BAKED PIE SHELL

1^1/4 cups all-purpose flour
^{1}/4 teaspoon salt
^{1}/3 cup shortening
3 to 4 tablespoons ice water

BLUEBERRY-CROWBERRY FILLING

^{1}/2 cup sugar
3 tablespoons cornstarch
^{1}/4 teaspoon salt
2 cups fresh blueberries or huckleberries
2 cups fresh crowberries (see Note)
1 to 2 tablespoons lemon juice
1 tablespoon butter

Whipped cream or sour cream, sweetened if desired
Grated lemon peel

To prepare the pie shell, combine the flour and salt in a bowl. Cut in the shortening with a pastry blender or rub together with your fingers until crumbly. Gradually sprinkle in the water, 1 tablespoon at a time, until the dough holds together when gathered into a ball. Knead lightly a few times. Flatten the dough into a 5-inch disk, dust with flour, and wrap in plastic. Chill for at least 30 minutes or up to 24 hours. If made ahead, let stand at room temperature for 30 minutes before rolling out.

Preheat the oven to 425 degrees.

Roll out the pastry disk on a lightly floured surface to a 12-inch circle. Fold into quarters and transfer to a 9-inch pie plate. Unfold and ease the pastry into the pan. Trim the crust to a $^1/_2$-inch overhang and flute the edge decoratively. Place a piece of foil over the crust and fill with pie weights or dried beans. Bake for 10 minutes. Remove the pie weights and foil and continue to bake for about 5 minutes more, or until golden. Cool completely on a rack.

To prepare the filling, combine the sugar, cornstarch, and salt in a saucepan. Combine the blueberries and crowberries. Stir in 2 cups of the berries and $^1/_4$ cup of water. Bring the mixture to a boil and cook for 1 minute. Reduce the heat and simmer, stirring constantly, until thick, about 5 minutes. Remove from the heat, stir in the lemon juice and butter, and let cool.

Place the remaining 2 cups of berries in the pie shell and spread the cooked berries over the top. Refrigerate for up to 4 hours before serving.

Serve with a dollop of whipped cream and a sprinkling of lemon peel.

Note: If crowberries are not available, increase the amount of blueberries or huckleberries to 4 cups.

Glazed Strawberry Pie

Makes one 9-inch pie

When I was a kid, my favorite going-out-to-dinner treat was a hamburger and a piece of "mile-high strawberry pie."

7 cups (about 2^1/$_2$ pounds) fresh strawberries
1 cup sugar, less if the berries are very sweet
3 tablespoons cornstarch
1 tablespoon lemon juice
1 baked and cooled 9-inch pastry shell (see Note)
Whipped cream (optional)

Rinse and pat dry the strawberries. Remove the hulls. Crush 2 cups of the strawberries in a large saucepan. Stir in about 2 tablespoons of water, depending on how juicy the berries are. Stir together the sugar and cornstarch, then mix into the crushed berries. Cook over medium heat, stirring frequently, until the mixture comes to a boil. Boil for 2 minutes, or until the sauce is thickened and clear. Remove immediately from the heat and pour into a large bowl. Stir in the lemon juice. Cool.

Arrange the remaining berries, stem ends down, in the pie shell and spoon the berry sauce over the fruit. Or gently stir the berries into the berry sauce and spoon into the pastry shell. Chill until set.

Cut into wedges and garnish with whipped cream, if desired.

Note: Use the recipe for the Baked Pie Shell (page 117), your own favorite pastry recipe, or a frozen baked crust.

Blueberry–Brown Butter Tart

Makes 8 to 10 servings

Seattle resident Rebecca Bolin first made this tart when she was pastry chef at Michael's in Santa Monica, California. It was the first thing to go in the oven in the morning, she says. "The aroma of the brown butter hitting the egg-sugar mixture was pure heaven."

You'll need a fluted 9- or 10-inch tart pan with a removable bottom for this and the next recipe.

PÂTE SUCRÉE

$1^1/2$ cups cake or pastry flour
$1/2$ cup sugar
10 tablespoons ($1^1/4$ sticks) chilled unsalted butter,
    cut into $1/4$-inch cubes
1 large egg yolk
1 tablespoon heavy cream

BROWN BUTTER FILLING

3 large eggs
1 cup sugar
1 teaspoon vanilla extract
$1/4$ cup and 1 tablespoon all-purpose flour
12 tablespoons ($1^1/2$ sticks) unsalted butter

3 cups fresh or frozen blueberries (see Note)
1 cup sugar

Powdered sugar
Whipped cream (optional)

To prepare the pâte sucrée, combine the flour and sugar in the large bowl of an electric mixer. Using the paddle, mix at medium-low speed, gradually adding the cubes of butter. When the mixture resembles cornmeal in texture, add the egg yolk and cream. Mix only until the dough comes together. Shape the mixture into a disk, wrap in plastic, and refrigerate for at least 30 minutes or up to 24 hours. If made several hours ahead, allow the pastry to stand at room temperature for 30 minutes before rolling.

Roll out the pastry on a lightly floured surface to a 12- to 13-inch circle about $1/8$ inch thick. Fold into quarters and transfer to a 9- or 10-inch fluted tart pan with removable bottom. Unfold and ease into the pan. Press so the sides are even. Trim off the excess. Chill for at least 30 minutes.

To prepare the brown butter filling, whisk together the eggs and sugar in a medium bowl until pale yellow. Stir in the vanilla and add the flour, whisking only until combined. Set aside. Melt the butter in a medium-size sauté pan just until golden brown (noisette stage, the color of hazelnuts). Add the butter to the egg mixture and whisk only until combined.

Preheat the oven to 350 degrees.

Place 1 cup of the blueberries in the pastry-lined tart pan. Pour the brown butter filling evenly over the berries. Bake for 30 minutes or until the crust is golden brown. Cool completely on a rack.

An hour or so before serving, poach the remaining blueberries. Combine the sugar and 1 cup water in a saucepan. Bring to a low boil. Add the remaining 2 cups of the blueberries and let stand 15 seconds for fresh berries, 30 seconds for frozen berries. Strain, reserving and refrigerating the syrup for another time, if desired. Pile the berries in the center of the tart and let them roll to the edges.

Just before serving, sprinkle with powdered sugar. Serve with whipped cream, if desired.

Note: To use frozen blueberries, rinse the frost off the berries under cold running water and blot dry.

Variation: Omit the blueberries. After baking, arrange concentric circles of fresh raspberries over the cooled tart. Brush the berries with a little warmed currant jelly.

Multiberry Frangipani Tart

Makes 8 to 12 servings

I created this tart once, when I came home with a half flat of assorted bramble berries from my nearby farmers' market. I arranged the berries in concentric circles over a rich almond filling, starting with the sweetest on the outside and the more acid in the center, using raspberries, Boysen berries, Marion berries, Tayberries, and Logan berries, in that order. Another time, in late spring, a friend and I designed a mosaic of strawberries, raspberries, and blueberries.

ONE-CRUST PÂTE BRISÉE

1 1/4 cups all-purpose flour

1/4 cup sugar

1/8 teaspoon salt

5 tablespoons chilled butter, cut into small pieces

3 tablespoons shortening

3 to 4 tablespoons ice water

ALMOND FILLING

1/3 cup (3 ounces) almond paste

3 tablespoons sugar

2 tablespoons butter

1 large egg

1 large egg yolk

2 teaspoons all-purpose flour

3 to 4 cups assorted berries, rinsed and patted dry as necessary
   (see above for suggestions)

GLAZE

1/3 cup seedless raspberry jam

2 teaspoons framboise, Kirsch, or water

To prepare the pâte brisée, combine the flour, sugar, and salt in a bowl. Cut in the butter and shortening with a pastry blender or rub together with your fingers until crumbly. Gradually sprinkle in the water, 1 tablespoon at a time, until the mixture holds together when gathered into a ball. Knead lightly a few times, dust with flour, and wrap in plastic. Chill for at least 1 hour or up to 24 hours. If made ahead, allow the pastry to stand at room temperature for 30 minutes before rolling.

To prepare the almond filling, combine the ingredients in a food processor and pulse on and off until smooth. Set aside.

Roll out the pastry on a lightly floured surface to an 12- or 13-inch circle. Fold into quarters and transfer to a 9- or 10-inch fluted tart pan with a removable bottom. Unfold and ease into the pan. Press so the sides are even. Trim off the excess. Chill for at least 15 minutes.

Preheat the oven to 400 degrees.

Pour the almond filling into the tart shell. Arrange the berries over the filling in an attractive design. Bake for 15 minutes. Reduce the heat to 300 degrees and continue to cook for 25 to 30 minutes longer, or until a knife inserted near the center comes out clean. Cool on a rack.

To prepare the glaze, combine the jam and framboise and heat in a microwave oven or a small saucepan until liquid. Remove the cooled tart from the pan. Place on a platter and brush with the glaze.

Note: The tart is best if served within a few hours of baking.

Free-Form Cranberry-Apple Tart

Makes one 10-inch tart

Don't worry if the crust isn't perfectly even; this country-style tart is supposed to be rustic-looking.

PASTRY

1 cup all-purpose flour
$1/4$ cup yellow cornmeal
2 teaspoons sugar
$1/4$ teaspoon salt
5 tablespoons chilled butter, cut into $1/2$-inch dice
2 tablespoons sour cream or yogurt
2 tablespoons ice water, or more if needed

CRANBERRY-APPLE FILLING

$1/4$ cup sugar
3 tablespoons all-purpose flour
$1/2$ teaspoon ground cinnamon
3 tablespoons chilled butter, cut into $1/2$-inch dice
4 or 5 cooking apples (about $2 1/2$ pounds), such as
    Jonagold, Elstar, Gravenstein, Newtown Pippin, or Idared
$1/2$ to $3/4$ cup dried cranberries

To prepare the pastry, place the flour, cornmeal, sugar, and salt in the bowl of a food processor. Pulse on and off until combined. Add the butter and toss with your fingers to coat the pieces with flour. Pulse on and off about 15 times, or until the butter is the size of small peas. Combine the yogurt and 2 tablespoons ice water. With the motor running, add the yogurt mixture all at once, then pulse on and off for about 10 seconds, stopping before the dough becomes a solid mass. The dough should hold together when pinched. Add a little more ice water, if necessary. Turn out onto a sheet of plastic wrap and press the dough together to form a 6-inch disk. Wrap and refrigerate for at least 1 hour or up to 2 days. If made ahead, let stand at room temperature 30 minutes before rolling out.

To prepare the filling, combine the sugar, flour, and cinnamon in a large bowl. With your fingertips, 2 knives, or a pastry blender, work in the butter until the mixture is crumbly and holds together in irregular lumps. Set aside.

Peel, core, and quarter the apples and cut each quarter crosswise into 4 or 5 chunks. You should have about 5 cups.

Preheat the oven to 450 degrees.

Roll out the pastry on a lightly floured surface to a 15-inch circle. Fold gently in quarters and transfer to a large baking sheet, such as a 14-inch air-pillow pan. Unfold the dough and arrange the apples in the center 10 inches of dough, leaving a $2^1/2$-inch border. Dot with dried cranberries. Sprinkle with the sugar mixture. Bring the outer edge of the dough $2^1/2$ inches over the apples, pleating as needed.

Bake for 25 or 30 minutes, or until the apples are tender. If necessary, cover with foil to prevent the crust from becoming too brown. Place the pan on a rack for 10 minutes, then transfer to a serving dish.

Serve warm or at room temperature.

Variation: Replace half of the dried cranberries with fresh or partially frozen cranberries.

Other Desserts

Marion Berry and Peach Cobbler

Makes about 6 servings

Last summer, we were touring the Sonoma wine country with my sister Patricia and her kids. The blackberries growing along the back country roads were at their most flavorful peak. With lots of pickers, we had enough for a cobbler in no time.

2 tablespoons sugar
2 teaspoons all-purpose flour
2^1/$_2$ to 2^3/$_4$ cups Marion berries or other blackberries
2 cups peeled and diced peaches or nectarines
1 teaspoon lemon juice

BISCUIT TOPPING

1 cup all-purpose flour
2 tablespoons sugar
1 teaspoon baking powder
^{1}/$_4$ teaspoon salt
4 tablespoons (^{1}/$_2$ stick) chilled butter
^{1}/$_4$ cup plus 1 teaspoon milk
Cinnamon sugar

Vanilla ice cream (optional)

To prepare the fruit, combine the sugar and flour in a large bowl. Add the berries, peaches, and lemon juice and toss well. Pour into a 2-quart casserole or baking dish and set aside.

Preheat the oven to 425 degrees.

To prepare the topping, combine the flour, sugar, baking powder, and salt in a bowl. Cut in the butter with a pastry blender or rub together with your fingers until crumbly. Add the ^{1}/$_4$ cup milk and stir just until the dough holds together.

Knead gently a few times. Roll out on a lightly floured surface to fit the top of the casserole. Cut into wedges and arrange over the fruit. Brush with the remaining 1 teaspoon milk and sprinkle with cinnamon sugar. Bake for 20 minutes. Reduce the heat to 325 degrees and continue to cook until the topping is brown and the fruit bubbles, about 20 minutes longer.

Serve warm or at room temperature, with a scoop of vanilla ice cream, if desired.

To Substitute Berries

The various sweet berries (strawberries, blueberries, and bramble berries) are generally considered to be interchangeable. But do check the sugar level.

Most strawberry recipes will work with raspberries, and vice-versa.

The bramble berry family will need an adjustment of the sweetener. Raspberries are usually the sweetest and Tayberries, the tartest.

Cranberry and red currant recipes are often interchangeable.

Raspberry Bread Pudding

Makes about 6 servings

My friend Odile Buchanan, an artist, has a wonderful palate. She first sampled this simple bread pudding while visiting her sister-in-law in Basel. She often uses raspberries in place of the traditional Italian plums.

> 3 large eggs
> $1/2$ cup plus $1^1/2$ teaspoons sugar
> 2 cups milk
> 4 cups ($3/4$-inch cubes) day-old French or country white bread,
>   crusts removed
> $3/4$ cup fresh or partially thawed frozen raspberries
> 1 tablespoon butter, cut into small pieces
> Unsweetened whipped cream, optional

Preheat oven to 400 degrees.

Beat the eggs with an electric mixer until foamy. Beat in the $1/2$ cup sugar. Add the milk and beat until mixed. Arrange the bread cubes in a buttered $1^1/2$-quart soufflé dish. Pour the egg mixture over the bread cubes. Place generous spoonfuls of berries in 6 or 7 places in the pudding. Push the bread down until it is moistened. Dot with butter and sprinkle with the remaining $1^1/2$ teaspoons sugar. Bake for 45 to 50 minutes, or until a table knife inserted near the center comes out clean.

Serve warm or at room temperature, with unsweetened whipped cream, if desired.

Note: If the baking dish is microwave-proof, the pudding can be rewarmed in a microwave oven.

Strawberries and Rhubarb with Biscuit Topping

Makes 8 servings

A winning combination of tangy rhubarb and sweet bits of strawberry. The trick is to slip the rhubarb into the oven to start cooking while you prepare the topping, in this case, tender little biscuits.

$2/3$ cup (packed) brown sugar
1 tablespoon cornstarch
6 cups sliced ($1/2$ inch thick) rhubarb

BISCUITS

$1^1/2$ cups all-purpose flour
$1/3$ cup sugar
$1^1/2$ teaspoons baking powder
$1/4$ teaspoon salt
3 tablespoons chilled butter, cut into small pieces
$1/2$ cup low-fat milk

4 cups strawberries, rinsed, hulled, and halved or quartered
Cinnamon sugar
Vanilla ice cream or frozen yogurt

Preheat the oven to 400 degrees.

Combine the brown sugar and cornstarch in a bowl. Add the rhubarb and toss to mix. Arrange in a 13 × 9 × 2-inch baking pan, and sprinkle with $1/4$ cup of water. Cover with foil and bake for about 15 minutes while preparing the biscuits.

To prepare the biscuits, combine the flour, sugar, baking powder, and salt in a bowl. Work in the butter with your fingertips or a pastry blender until crumbly. Stir in the milk with a fork to make a soft dough, mixing just until moistened. Gently form the dough into 8 small rounds a scant $1/2$ inch thick.

Remove the rhubarb from the oven and stir in the strawberries. Arrange the biscuits on top and sprinkle with cinnamon sugar. Return to the oven and bake, uncovered, for about 20 minutes, or until the biscuits are light brown and the fruit bubbles.

Serve warm or cool with vanilla ice cream or frozen yogurt.

Midsummer Eve Strawberry Soup

Makes 6 servings

Save a few of the prettiest berries for the garnish. Serve the soup as a first course or as a tart-sweet dessert. On a hot day, you might want to chill the soup plates. And for a special occasion, a Champagne accompaniment would be nice.

3 cups fresh strawberries, rinsed and hulled, or frozen strawberries
2 tablespoons sugar
1 1/2 teaspoons cornstarch
2 cups cranberry juice cocktail
2 teaspoons red wine vinegar
1 cup buttermilk
3 to 6 strawberries, whole or halved, for garnish

Puree the berries in a food processor, pulsing on and off until smooth. Combine the sugar and cornstarch in a nonreactive saucepan. Stir in the cranberry juice and vinegar. Cook over medium heat until the mixture bubbles and is clear. Add the pureed berries and cook for 2 minutes, stirring constantly. Remove from the heat and strain into a bowl, pressing firmly. Discard the pulp and seeds.

Serve when cool or cover and refrigerate until chilled completely, about 2 hours. Stir in the buttermilk. Pour into bowls. Garnish with the remaining strawberries and/or drizzle a few drops of buttermilk onto each serving.

Berry Crunch

Makes 8 to 10 servings

Evette Hackman, Food and Nutrition professor at Seattle Pacific University, adds two unusual ingredients—jalapeño hot pepper jelly and Tabasco sauce—to her low-fat dessert. She suggests topping the crunch with low-fat ice cream.

3 cups fresh or frozen blueberries, rinsed, blotted dry, and picked
   over if necessary
3 cups fresh or frozen red raspberries, rinsed, blotted dry, and
   picked over if necessary
$1/2$ cup plus $1/4$ cup (packed) brown sugar
1 tablespoon hot or jalapeño pepper jelly
$1/4$ to $3/4$ teaspoon Tabasco sauce
1 tablespoon lemon juice
$1^1/2$ cups quick or regular rolled oats
$1/2$ teaspoon ground cinnamon
$1/4$ teaspoon ground cardamom
4 tablespoons ($1/2$ stick) melted butter
1 teaspoon vanilla extract
Vanilla ice cream

Preheat the oven to 350 degrees. Coat a 11 × 7 × 2-inch pan with vegetable oil cooking spray or lightly coat with butter.

Combine the berries and arrange $5^1/2$ cups in the pan. Toss the remaining $1/2$ cup of berries with the $1/4$ cup of brown sugar, the pepper jelly, Tabasco, and lemon juice in a small saucepan. Cook and stir over medium heat until heated through. Pour over the berries. Combine the remaining $1/2$ cup brown sugar and the oats, cinnamon, cardamom, butter, and vanilla in a bowl. Spread evenly over the berries. Bake until the berries begin to bubble, about 30 minutes for fresh berries, 35 to 40 minutes for frozen berries.

Serve warm or cool with vanilla ice cream.

Gooseberry Fool

Makes about 6 servings

Sheffield-born Barbara Haney often makes this favorite English dessert in the summer with gooseberries from her garden. She recommends "topping and tailing," that is, removing the stems and blossom ends, with scissors. Of course, you can use your fingers.

> 12 ounces (heaping 2 cups) fresh gooseberries (see Note)
> 1¹/₂ tablespoons butter
> 2 to 3 tablespoons sugar, or to taste
> 1 cup chilled whipping cream

Remove the stems and blossom ends of the gooseberries. Melt the butter in a heavy saucepan. Add the gooseberries and 2 tablespoons of sugar. Cook, covered, over low heat until soft and mushy, 15 to 25 minutes.

Crush the berries thoroughly with a fork. If the skins are particularly tough, place the berry mixture in the bowl of a food processor and pulse on and off, taking care that the mixture retains some texture. Cool completely. Correct the amount of sugar, if necessary. The mixture should be tart, however.

Whip the cream to soft peaks and fold in the gooseberry mixture. Spoon into a serving bowl, individual dishes, or parfait glasses. Chill for at least 1 hour before serving.

Note: One 16¹/₂-ounce can of gooseberries, drained, may be substituted for the fresh gooseberries. Cook for only 5 to 10 minutes.

Fresh Peach Melba

Makes 6 servings

When Auguste Escoffier created this dish for the famous 19th-century Australian diva, Nellie Melba, he brought the classic combination of raspberries and peaches to new heights.

3/4 cup sugar
1 teaspoon vanilla extract
3 large firm-ripe peaches, peeled, halved, and pitted
1 package (10 ounces) thawed frozen
   sweetened raspberries, undrained (see Note)
1 tablespoon cornstarch
1 to 3 teaspoons lemon juice
Vanilla ice cream

Combine 1 cup of water, the sugar, and vanilla in a large saucepan. Bring to a boil. Add the peach halves to the syrup. Reduce the heat and simmer, turning the peaches occasionally, for 10 minutes, or until barely tender. Remove from the heat and transfer to a bowl. Place a saucer on top to keep the peaches submerged. Refrigerate for up to 24 hours.

Puree the raspberries and their liquid in a food processor. Strain to remove the seeds. Combine the cornstarch and 1 tablespoon of water in a saucepan. Stir in the berry puree. Cook on low heat for 4 to 5 minutes, or until clear and thickened. Stir in lemon juice to taste. Transfer to a bowl and cool.

To assemble, place a drained peach half, cut side up, in each of 6 dishes. Add a scoop of vanilla ice cream and pour the Melba sauce over the top.

Note: To use fresh berries instead of frozen, substitute 1 1/2 cups (tightly packed) fresh raspberries with 1/4 cup sugar.

Chocolate-Dipped Strawberries

Makes about 32 chocolate-dipped strawberries

Fran Bigelow, of Fran's Chocolates in Seattle, suggests using Callebaut chocolate for dipping fruit, but any good-quality chocolate will do.

6 ounces bittersweet or semisweet chocolate, cut into small pieces
32 firm medium-size strawberries

Place the chocolate in the top of a double boiler or in a small bowl and set it over a pan of barely simmering water. Heat until the chocolate melts, stirring occasionally. Don't rush, and be careful not to get any water in the chocolate.

Rinse the berries and pat dry. Dip each berry in the chocolate, coating two thirds of the berry. Let any extra chocolate drip off. Place the berries on wax paper–lined baking sheets. Refrigerate for about 15 minutes, or until the chocolate is set.

To serve, loosen the berries from the wax paper and arrange on a platter. Coated strawberries will keep for an hour or two at room temperature.

Fresh Cranberry Sorbet

Makes 6 to 8 servings

This homemade sorbet makes a refreshing palate cleanser at Thanksgiving. If you have a bag of cranberries in your freezer, it makes a pleasant summer dessert.

3^1/2 cups (12-ounce bag) fresh or frozen cranberries,
    rinsed and picked over
1^1/2 cups sugar
^{3}/4 cup orange juice
2 tablespoons fresh lemon juice
1 teaspoon grated orange peel

Combine the cranberries, sugar, and 2^1/2 cups of water in a large saucepan. Bring to a boil, then reduce the heat and simmer for 10 minutes, stirring occasionally. Transfer to a large bowl and let cool.

Place the mixture in a food processor or blender and pulse on and off until thoroughly pureed. Strain through a sieve and discard the skins. Transfer the mixture to a bowl. Add the orange juice, lemon juice, and orange peel and mix well. Chill thoroughly. Freeze in an ice-cream maker according to the manufacturer's directions.

Strawberry Margarita Sorbet

Makes about 3 cups

Wow! This tastes like a real margarita.

> ³/4 cup sugar
> 3¹/4 cups strawberries, rinsed and hulled
> ¹/4 cup fresh lime juice
> ¹/4 cup tequila or rum
> 1 teaspoon Triple Sec (optional)
> Strawberry fans, for garnish
> Lime slices, for garnish

Combine the sugar and ¹/2 cup of water in a saucepan. Bring to a boil, stirring occasionally, until the sugar dissolves. Transfer to a 1-quart measure and cool completely.

Place the mixture in a food processor or blender and pulse on and off until pureed. Add the strawberries to the cooled sugar syrup. Add the lime juice, tequila, and Triple Sec, if desired. Stir well. Chill thoroughly. Freeze in an ice-cream maker according to the manufacturer's directions.

Spoon into stemmed glasses. Garnish with strawberry fans and lime slices. Serve immediately.

Index